INSTITUTIONAL CHALLENGES TO INTERMODAL TRANSPORT AND LOGISTICS

Jason Monios offers a compelling approach to the concept of port regionalisation and offers convincing case studies underlining the need for strong involvement by terminal operators as a key success factor for inland terminal development. In areas of overlapping hinterland it illustrates the need for collective action between seaports. Overall an interesting perspective to the development of hinterland logistics.
Eduard de Visser, Director Strategy & Innovation, Port of Amsterdam

Transport and Mobility Series

Series Editors: Richard Knowles, University of Salford, UK and Markus Hesse, Université du Luxembourg and on behalf of the Royal Geographical Society (with the Institute of British Geographers) Transport Geography Research Group (TGRG).

The inception of this series marks a major resurgence of geographical research into transport and mobility. Reflecting the dynamic relationships between socio-spatial behaviour and change, it acts as a forum for cutting-edge research into transport and mobility, and for innovative and decisive debates on the formulation and repercussions of transport policy making.

Also in the series

Institutional Challenges to Intermodal Transport and Logistics

Governance in Port Regionalisation and Hinterland Integration

JASON MONIOS
Edinburgh Napier University, UK

Routledge
Taylor & Francis Group

LONDON AND NEW YORK

First published 2014 by Ashgate Publishing

Published 2016 by Routledge
2 Park Square, Milton Park, Abingdon, Oxon OX14 4RN
711 Third Avenue, New York, NY 10017, USA

First issued in paperback 2018

Routledge is an imprint of the Taylor & Francis Group, an informa business

British Library Cataloguing in Publication Data
A catalogue record for this book is available from the British Library

The Library of Congress has cataloged the printed edition as follows:
Monios, Jason.
  Institutional challenges to intermodal transport and logistics : governance in port regionalisation and hinterland integration / by Jason Monios.
    pages cm—(Transport and mobility)
  Includes bibliographical references and index.
  ISBN 978-1-4724-2321-4 (hardback)
1. Containerization. 2. Freight and freightage. 3. Business logistics. 4. Harbors--Economic aspects. I. Title. II. Series: Transport and mobility series.
  TA1215.M66 2014
  385'.72--dc23

2014003393

ISBN 13: 978-1-138-54664-6 (pbk)
ISBN 13: 978-1-4724-2321-4 (hbk)

i.m. Ian Mathie

Contents

List of Figures and Maps

Figures

Maps

List of Tables

Note on the Author

Dr Jason Monios is a Senior Research Fellow at the Transport Research Institute, Edinburgh Napier University, UK, where he leads research and consultancy projects in the Maritime Transport & Logistics Research Group. His primary research area is intermodal freight transport, with a specific interest in the relations between ports and inland actors in managing hinterland access.

Preface

This book is part of a stream of research on intermodal transport and logistics carried out at the Transport Research Institute, Edinburgh Napier University from 2009 to 2012. While much research has been conducted on the increasing vertical and horizontal integration in the maritime transport sector, the interest of this research was to examine how such processes have proceeded in inland space, where freight flows are inherently more complex and thus resistant to such influences.

While freight transport has risen in prominence in the last decade as a specialism within transport geography, it remains under-researched. Certainly, it remains under-conceptualised. If transport geography has tended in the past to be less theoretical than other areas of geography (Hall et al., 2006), freight transport has likewise been treated less from theoretical perspectives than other kinds of transport (Ng, 2013). Transport geography is also less interdisciplinary, at times too one-dimensional, perhaps overlooking the fact that transport 'is not just about modes and movement but also about politics, money, people and power, and there is a need for transport geography to be a more *human* geography' (Shaw and Sidaway, 2010: p.515). For freight transport, this means that it has been insufficiently addressed within its logistics context.

While this book does provide an overview of practical issues in the industry, space limits preclude a full examination of the practical and operational aspects of freight transport. This book is not a practical manual, nor is it a textbook. Readers seeking a sustained treatment of the practical aspects of intermodal freight transport are directed to Lowe (2005). For a book-length treatment of the role of transport in logistics and distribution, there is Hesse (2008). Other works take a wider view on the historical development of containerised transport (Levinson, 2006; Bonacich and Wilson, 2008), or globalisation (Dicken, 2011; Friedman, 2005). Similarly, those desiring a textbook of transport geography should look to Knowles et al. (2008) and Rodrigue et al. (2013). This book is an analysis of some of the more conceptual academic issues, nonetheless firmly rooted in real world examples.

The topic of intermodal transport has risen to prominence in the last five years or so, with an explosion of academic papers on intermodal terminals and dry ports appearing in a range of academic journals,[1] aligned with much public funding for research that is hoped to support government goals of emissions reduction,

---

1 An early book-length publication on intermodal transport was an edited collection of pieces by several authors (Konings et al., 2008).

congestion amelioration and economic growth. As this book will attempt to show, such motivations are not easy to achieve, especially all at the same time. The aim of this book was therefore to draw together the subject of intermodal transport within its logistics context, in order to demonstrate that the difficulties challenging success are not simply practical but due to the organisational complexities of the sector. While the maritime transport sector has witnessed a wave of consolidation via mergers and acquisitions, such integration has not occurred in land transport, and care should be taken assuming intermodal transport by land is as straightforward as a movement between two large port terminals operated by the same globalised and vertically integrated terminal operator and shipping line. Operational models of landside operators are crucial in terms of whether an operator can succeed in developing intermodal services, based on the ability to cooperate, integrate, consolidate and plan. These are not just operational concerns but are in many instances derived from the governance model. If policy makers are to achieve their goal to promote intermodal transport, public sector planners and funders require an understanding not just of potential cost and emissions savings based on an ideal scenario of regular intermodal traffic flows but of the institutional difficulties in consolidating and planning regular balanced intermodal shuttles.

In order to address these issues, an interdisciplinary approach has been taken, thus addressing another aim which was to expand current theoretical approaches to transport geography, applying insights from economic geography and institutional economics. This approach reflects my belief that transport geography, in particular the geography of freight transport, has much to learn from, and contribute to, other sub-disciplines of geography. The geography of intermodal transport and logistics provides fertile ground for exploring a variety of theoretical propositions, anchored on an emerging conceptualisation of the institutional structures underpinning successful intermodal transport and logistics. In particular, port regionalisation and other strategies of hinterland integration provide an illuminating canvas on which to trace conceptual developments that can then be used to inform other areas of theoretical endeavour, whether spatial (e.g. analyses of concentration and centralisation) or institutional (e.g. analyses of governance and institutional transformation).

It should also be noted here that the book is concerned with rail rather than water transport. While intermodal transport includes container movements by both rail and barge, rail transport is by far the most common topic in the literature. The barge literature tends to focus on operations rather than strategic development and the institutional aspects of port integration.

Acknowledgements

I would like first to thank over 100 interviewees who generously shared their time and knowledge with me during the course of this research. Without their assistance this book would not have been possible. I would particularly like to thank those who helped arrange the interviews: Xavier Gesé (Spain), Dr Fedele Iannone and Angelo Aulicino (Italy), Gavin Roser (UK) and Bruce Lambert (USA).

Funding for the European research in this book was provided by the Dryport project, funded by the European Union through the Interreg IVB North Sea Region programme. The American research was funded by a grant from the Royal Society of Edinburgh.

This research would not have been possible without Dr Gordon Wilmsmeier, with whom I collaborated on many of the ideas developed in this book. In addition, the director of my institute Professor Kevin Cullinane has been supportive throughout this project.

Finally, my thanks to the series editors, Professor Markus Hesse and Professor Richard Knowles, for the offer to publish in this series, as well as guidance on the proposal and final manuscript.

Jason Monios

List of Abbreviations

3PL	Third-party logistics provider
AAR	American Association of Railroads
AASHTO	American Association of State Highway Transportation Officials
ARC	Appalachian Regional Commission
BNSF	Burlington Northern Santa Fe
CNRS	Company neutral revenue support
DC	Distribution centre
DFT	Department for Transport
DOT	Department of Transport
DRS	Direct Rail Services
ECML	East coast mainline
ECT	European Container Terminals
EFLHD	Eastern Federal Lands Highway Division
FCL	Full container load
FFG	Freight Facilities Grant
FGP	Factory gate pricing
FHA	Federal Highways Association
HAR	Hinterland Access Regime
HMT	Harbour Maintenance Tax
ILU	Intermodal loading unit
ISTEA	Intermodal Surface Transportation Efficiency Act
ITU	Intermodal transport unit
LCL	Less than container load
MarAd	Maritime Administration
MPO	Metropolitan Planning Organisation
MSRS	Modal Shift Revenue Support
MTO	Multimodal transport operator
NDC	National distribution centre
NIE	New institutional economics
NS	Norfolk Southern
ORDC	Ohio Rail Development Commission
PCC	Primary consolidation centre
PPP	Public-private partnership
RDC	Regional distribution centre
REPS	Rail environmental benefit procurement scheme
RTI	Nick J. Rahall Appalachian Transportation Institute
RTP	Regional transport partnership

SAFETEA-LU	Safe, Accountable, Flexible and Efficient Transportation Act: A Legacy for Users
SFN	Strategic freight network
SKU	Stock keeping unit
TAG	Track access grant
TEA-21	Transportation Equity Act for the 21st Century
TEU	Twenty-foot equivalent unit
TIGER	Transportation Investment Generating Economic Recovery
UP	Union Pacific
VDRPT	Virginia Department of Rail and Public Transportation
WCML	West coast mainline
WFG	Waterborne freight grant

The term 'theorist' is often applied to those who deal mainly in abstractions and abjure empirical verification, rather than to those who take up knotty problems, hypothesize about their nature and causality, and marshal evidence in support of their views.[1]

Markusen, 2003: p.704

A scientific discipline without a large number of thoroughly-executed case studies is a discipline without systematic production of exemplars, and a discipline without exemplars is an ineffective one.[2]

Flyvbjerg, 2006: p.219

1   From 'Fuzzy Concepts, Scanty Evidence, Policy Distance: The Case for Rigour and Policy Relevance in Critical Regional Studies' by Ann Markusen, p.704, *Regional Studies* 33 (9), 2003, copyright © Regional Studies Association, reprinted by permission of Taylor & Francis Ltd, www.tandfonline.com on behalf of The Regional Studies Association.

2   From 'Five Misunderstandings About Case-Study Research' by Bent Flyvbjerg, *Qualitative Inquiry* 12 (2) p.219, copyright © 2006 by SAGE Publications. Reprinted by Permission of SAGE Publications.

Chapter 1

Introduction

Why Intermodal Transport?

The subject of intermodal transport is raised with increasing frequency in government policy and planning documents, industry promotional brochures and academic journal articles. It is promoted by governments as a way to reduce emissions through modal shift as well as support economic growth through reduced congestion and better access to global trade routes. Terms such as 'dry ports' are used in public planning documents as though they were panaceas for uncompetitive regions and the transport strategy of every public authority seems to involve some kind of logistics platform to consolidate business, lower transport costs and boost local employment. Firms promote themselves as integrated logistics providers offering intermodal options, and the buzzwords of intermodal, co-modal and synchromodal strive for dominance on marketing material. Meanwhile, academics try to keep up with these trends, measuring the transport costs, time savings or emissions reductions claimed by new developments.

In Europe, despite large public investments, intermodal transport continues to struggle against road haulage, due to various operational reasons such as short distances, fragmented demand and equipment imbalances. While these issues also challenge short-distance intermodal traffic in the United States, the vast physical geography of the country means that there is a large market for intermodal traffic on long-distance routes. Developing economies are catching up, although the requirements are more for improved logistics than emissions reductions or even lower transport costs.

Intermodal transport strategies can be based on differing motivations, actors, functions and logistics models. Intermodal terminals can be close to the port, mid-range or distant. The goal of the port may be to ease congestion or to capture hinterlands as instruments of port competition. Terminals and corridors can be developed by port authorities, port terminal operators and transport providers such as rail operators or third-party logistics providers (3PLs), or they can be developed by public bodies, whether national, regional or local. Business models can be based on economies of scale on high-capacity, long-distance links or on logistics improvements such as providing containerisation facilities or allowing fast-track customs clearance. Even where the operational challenges of intermodal transport have been examined, sufficient effort has not always been made to disentangle the different elements. It is argued here that if the institutional relationships underpinning each development are not understood then the development may well fail, even if sufficient demand is known to exist.

This lack of clarity regarding the business strategy of an intermodal transport development is partly because, while the operational realities of intermodal transport are relatively well known, the institutional challenges are less well understood. Institutional approaches have been applied in other areas of geography and other disciplines such as economics. It is only in recent years that such a lens has been turned on the geography of freight transport, and this has only taken the form of a small number of journal articles. This work represents the first book-length treatment of the topic.

Intermodal Transport and Logistics in the Context of Port Regionalisation

Traditional spatial analyses of freight transport have been expanded to cover institutional relationships governing the complexity of transport connections. Recent research has shown how spatial development is to a large degree an institutional problem (de Langen and Chouly, 2004; Van der Horst and de Langen, 2008; Van der Horst and Van der Lugt, 2009; Van der Horst and Van der Lugt, 2011). This development recognises that modern transport actors operate in an increasingly complex and sophisticated transport and logistics environment, embedded within multi-scalar planning regimes. The notion of transport solely as a derived demand has been challenged and reformulated as an integrated demand (Hesse and Rodrigue, 2004; Rodrigue, 2006; Panayides, 2006; Hesse, 2008). As such, the relationship between goods flows and spatial development is complicated by networks of nodes and corridors that may not perform their key functions adequately, potentially constrained not just by physical infrastructure deficits but a lack of connectivity or an inability to fit into wider networks. The focus of this research is on the use of rail transport;[1] a firm's decision to shift to this mode can be driven by many factors, such as external pressures (e.g. fuel price, legislation, customer pressure) or logistics strategy (e.g. central warehouse or distributed network, private fleet or 3PL) (Eng-Larsson and Kohn, 2012). Yet, according to some authors, the role of transport in logistics and the broader field of supply chain management has been under researched (Mason et al., 2007).

This book addresses the institutional challenges to intermodal transport and logistics, through the frame of port regionalisation (Notteboom and Rodrigue, 2005), which is an approach to port development that focuses on the inland aspects of the process, as well as taking port development models from a spatial focus to a focus on institutions. The research identifies and examines the processes implicit within the port regionalisation concept; namely, intermodal

---

1   While intermodal transport includes container movements by both rail and barge, rail transport is by far the most common topic in the literature. The barge literature (e.g. Choong et al., 2002; Trip and Bontekoning, 2002; Groothedde et al., 2005; Konings, 2007; Konings et al., 2013) tends to focus on operations rather than terminal development and the institutional aspects of port integration.

terminal development, logistics integration strategies and the institutional processes of resolving collective action problems. While port regionalisation covers more than intermodal transport, these three processes are all inextricably linked with the concept because the relevant infrastructure and the integration of dominant industry players provide means to capture or control key corridors and load centres. The port regionalisation concept relies on effective intermodal transport infrastructure and operations underpinning the levels of integration required for hinterland capture and control. Port regionalisation cannot, therefore, be fully understood (and hence theorised) without greater analysis of the key issues arising from an in-depth analysis of the spatial and institutional characteristics of intermodal transport and logistics.

The fundamental concern of this book is to understand the process of transport and logistics integration between ports and hinterlands in comparison to that previously observed in sea transport. In the past decade, shipping lines and port terminal operators have consolidated and integrated their portfolios through mergers and acquisitions, resulting in a small number of dominant firms. These firms have since benefited from significant economies of scale and scope and enabled them to provide something of a seamless intermodal transport movement from port to port. Many actors are endeavouring to pursue this same trend inland, but for true intermodality to be successful and economically feasible, land transport operations require a similar level of consolidation to that of the sea leg. The different characteristics of land transport, however, challenge this goal (McCalla et al., 2004). As Graham (1998: p.135) wrote: 'the land-side is characterized by relatively low investment, high operating expenses, little scale incentive to collective operation and a considerable level of unremunerated activity requiring cross payment out of sea freight'. This book investigates the nature of these processes in order to identify the institutional barriers to intermodal transport and logistics.

The Research Approach

The research in this book follows a predominantly inductive approach. Inductive reasoning commonly proceeds from the specific to the general (as opposed to deductive approaches which flow in the opposite direction). It begins with observations then identifies patterns, from which hypotheses and theory may be developed. Like deductive approaches, inductive research may also begin with theory from the literature; however, while deductive research aims to test a theory (or to test hypotheses derived from a theory), in this case the inductive approach looks for aspects that are not explained by the current theory, and attempts to develop new theory that can account for these gaps.

Kelle (1997: unpaginated) suggested that 'the theoretical knowledge of the qualitative researcher does not represent a fully coherent network of explicit propositions from which precisely formulated and empirically testable statements can be deduced. Rather it forms a loosely connected "heuristic framework" of concepts which helps the researcher focus his or her attention on certain phenomena in the field'. The generalisation remains a working hypothesis or extrapolation rather than a grand theory, based on logic rather than statistics or probability (Cronbach, 1975; Patton, 2002). According to Seale (1999: p.52), the researcher is 'seeking for evidence within a fallibilistic framework that at no point claims ultimate truth, but regards claims as always subject to possible revision by new evidence'. He took this view forward by focusing on the skills of the individual researcher to construct a valid argument based on observable and presentable data. Seale argued that it is possible to follow a middle ground between positivist truth and socially constructed knowledge by remaining cognisant of the constructed nature of research even as this imperfect edifice is utilised to investigate a subject.

According to Yin (2009), a case study approach is appropriate when 'how' or 'why' questions are being asked, when the investigator does not have control over events (as one might in an experimental methodology) and when the phenomenon being studied cannot be separated from its context. Since all of these criteria are present in the current research, a case study methodology has been adopted, with differing approaches to the case study format in each of the three empirical chapters addressing each of the three research questions (see Chapter 2 for the derivation of the research questions).

Case studies can be based on both quantitative and qualitative data and indeed combined (Mangan et al., 2004; Näslund, 2002; Woo et al., 2011b). Qualitative case studies capture rich data and derive explanations from this 'thick description' (Stake, 1995). According to Merriam (1988: p.16), case studies are 'particularistic, descriptive, and heuristic and rely heavily on inductive reasoning in handling multiple data sources'. Therefore the key aspects of the case study approach are the depth available from the qualitative data, the particularity of each case, the fact that it is situated within its context and the attempt to understand the phenomenon from multiple perspectives (Simons, 2009).

The data collection for this research was based primarily on expert interviews; more than 100 subjects were interviewed for the empirical sections of this book. These data were supplemented by document analysis in order to build robust case studies that could then be analysed through thematic matrix construction. The case study approach has been adopted because it provides the rich data that is required to achieve the inductive aims of this research. The port regionalisation concept requires greater disaggregation, thus it is only after the analysis of detailed case studies that explanations can be offered and theories developed that can improve upon and refine the port regionalisation concept. The book does not have a spatial focus but is guided by theory. Thus the research questions are open-ended because the goal is not to apply findings to a specific spatial context (e.g. port regionalisation in the UK), but to generalise to theory (e.g. how can the theory be improved or expanded). A case study methodology is suitable for these inductive aims.

Chapter 4 follows a multiple-case design, analysing numerous inland terminal developments in Europe. Chapters 5 and 6 each utilise a single case design, taking a single case in depth in order to explore in rich detail how these issues play out in industry. The cases have been selected through purposive sampling (as opposed to random sampling utilised for a survey methodology), chosen to represent certain characteristics. In particular, the cases have been chosen primarily for theoretical purposes, as they are guided by their potential contribution to theory. They can be used to test current categories, explore new categories and refine them. This is particularly the case in Chapter 4, which is based on a multiple-case design, as each case contributes to the emerging classification of inland terminals. Chapters 5 and 6 are based on in-depth analysis of single cases; therefore, while those cases are relevant for theory, they have also been chosen partly because of their representative nature.

In presenting the research findings, it is important to retain the rich data as well as the summarised evidence for answering the research questions. Flyvbjerg (2006) argued that 'case studies … can neither be briefly recounted nor summarised in a few main results. The case story is itself the result. … the payback is meant to be a sensitivity to the issues at hand that cannot be obtained from theory' (pp.238–9). Furthermore, 'the problems in summarising case studies … are due more often to the properties of the reality studied than to the case study as a research method' (p.241). On the other hand, Miles and Huberman (1994) recommend regular use of tables and matrices both for guiding the analysis and presenting the findings.

The case studies in this book are therefore presented via a combination of narrative form and tabulated summaries of key evidence for the research factors, which are then discussed at the end of each chapter. Sub-factors are used to structure the findings but the overall research questions are to learn about how the process takes place, therefore a narrative style can discuss those issues and allows for more detail on actual practice. This approach enables readers to follow the research process and draw their own conclusions on its internal validity, as well as the external validity, or degree to which the findings are transferable to other cases (Seale, 1999; LeCompte and Goetz, 1982).

It is also important not to lose the link to the interview context, as the value of expert interviews is that they provide an insight into actual practice, which should not be subsumed beneath overly abstract categories. Therefore while the key findings have been summarised in tables presented throughout each chapter, they have been supplemented by the inclusion of examples of practice drawn from the interviews. Due to commercial sensitivity the detail has been kept fairly general where necessary (mostly in Chapter 5).

Generalising from case studies can be problematic. Bryman (2008: p.55) stated that generalisability or external validity is not the aim: 'case study researchers … do not think that a case study is a sample of one'. Case studies can be used to generalise to theoretical propositions rather than samples, meaning that the aim is to 'expand and generalise theories (analytic generalisation) and not to enumerate frequencies (statistical generalisation)' (Yin, 2009: p.15). Hammersley (1992) named these two kinds of generalisation theoretical and empirical and Lewis and Ritchie (2003) split statistical or empirical generalisability into two kinds, representational and inferential. In this research the goal is analytic generalisability, or generalising to theoretical propositions.

It is recognised that additional case studies will be required to continue this line of theoretical enquiry and strengthen the validity of the explanations offered in this book. On the other hand, one of the strengths of this research is the access to high calibre interviewees, particularly in Chapters 5 and 6. Therefore the case study methodology based primarily on semi-structured interviews offered the opportunity to capture valuable industry knowledge that would not have been possible through a different research design.

Structure of the Book

This book provides an overview of the important issues relating to intermodal transport and logistics, including the policy background, emerging industry trends and academic approaches. The three key features of intermodal transport geography will be established as intermodal terminals, logistics and corridors. An interdisciplinary perspective will be introduced, based on a discussion of the role of institutions, how they relate to transport and how they can enable understanding of some of the intractable challenges facing successful intermodal transport and logistics.

Chapter 2 provides an overview of intermodal transport and logistics, beginning with a brief history and then introducing the key actors and issues, examining the motivations of industry and government in reducing transport costs as well as decreasing emissions. Chapter 3 establishes the institutional approach that underpins the research in this book, charting key governance relationships between public and private sector actors.

Chapters 4–6 provide the empirical content via three detailed case study chapters based on extensive field work in Europe and the United States, exploring the theoretical and practical issues in real-world examples. Chapter 4 compares a sample of 11 European intermodal terminals, divided into one group with port investment and one group without, in order to identify different models of inland terminal development and improve inland terminal classifications that can then contribute to the port regionalisation concept, including classifications such as 'dry port', 'extended gate' and 'freight village'. Their key features and functions are discussed, including the roles of the public and private sectors, relations with ports and rail operators and their situation within the logistics sector. Chapter 5 analyses a case study of intermodal logistics in the UK. Large retailers are the primary drivers of intermodal transport in the UK, and they are explored in the context of their relationships with rail operators and 3PLs. Chapter 6 examines the development of an intermodal corridor in the United States, offering the opportunity to study a collective action problem in detail, in which several actors need to come together to solve a joint problem. Collective action is an arena where various actors can be influential due to the role of informal networking in managing freight corridors;

however, institutional constraints such as a conflict between legitimacy and agency and the limitations of institutional design restrict their ability to act directly.

Following the case study chapters, Chapter 7 uses the institutional literature from Chapter 3 to expand the empirical findings via a wider discussion of governance in transport and logistics, comparing the operational models identified in this research with other cases from the literature. In Chapter 8, a brief global comparison of intermodal freight transport and logistics covers Europe, North America, Asia, Africa and Latin America, explaining how the key trends and issues derive from the transport and logistics environments of each continent. For instance, differences exist between developed economies with mature transport and logistics industries and developing countries seeking access to more basic facilities such as inland customs clearance and sources of maritime containers.

Discussion of the challenges to intermodal transport and logistics demonstrates how a lack of integration and competing port and inland strategies challenge the necessary process integration that underpins successful intermodal transport. The empirical cases in this book reflect the differences of strategy and requirements in each area. The book concludes with some recent work on institutional adaptations as freight actors change in order to pursue the greater integration required for successful intermodal transport and logistics, establishing the main topics for future research.

Impact and Relevance of the Findings

Chapter 4 shows that ports can actively develop inland terminals, and differences exist between those developed by port authorities and those developed by port terminal operators. Furthermore, differences can be observed between those developed by ports and those developed by inland actors. The difficulties of successful terminal development unless embedded firmly within a rail, port or logistics model are demonstrated.

Chapter 5 demonstrates how the efficiency of rail freight services is challenged by the need to combine port and domestic movements which have different product, route and equipment characteristics. As large retailers are the primary drivers of intermodal transport in the UK, lessons can be learned for promoting intermodal transport in comparable contexts such as Europe where similar operational conditions prevail. The findings reveal that, while rail remains a marginal business, while the industry remains fragmented, while third-party consolidation is not pursued and while fragile government subsidy is still the basis of many flows, intermodal corridors cannot become instruments of hinterland capture and control for UK ports. The integration processes predicted by the port regionalisation concept cannot develop significantly unless the inland freight system is more integrated.

In Chapter 6, an interdisciplinary theoretical model is developed from the literature, attempting to unite institutional economics and economic geography. The literature suggests a conflict between legitimacy and agency and a limitation of political organisations due to their design, both of which are confirmed in the findings. These issues account for the high incidence of policy churn, lack of agency and, sometimes, lack of communication between the public and private sectors. The role of informal networking is found to be important as it can overcome institutional inertia, although it is difficult to capture this process, and harder still to attempt to institute it in another setting through policy action. The case study demonstrates how the institutional setting of intermodal corridors is changing through the influence of public-private partnerships on government policy. A reconciliation is identified between top-down planning approaches and bottom-up market-led approaches. Findings from Chapter 6 show that institutional design will continue to constrain integration between maritime and inland transport systems, suggesting that port regionalisation processes may not develop as the concept assumes. Legitimacy and agency are a problem for these organisations and if an infrastructure for collective action is not in place (and it is usually predominately a public infrastructure for collective action), then private firms will not act, thus challenging attempts at port regionalisation and keeping the maritime and inland spaces separate.

The governance framework developed in Chapter 7 highlights the key institutional relationships requiring further research, and when these findings are returned to the theoretical context of port regionalisation and hinterland integration, challenges to the integration required for successful intermodal transport are identified. A particular challenge is the difficulty in developing intermodal freight transport by the public sector while operational models do not align with transport demand, questioning the ability of public policy to support the industry. Thus more work is required on the correct approach in government policy, requiring an understanding not just of potential cost and emissions savings based on regular intermodal traffic flows but the institutional difficulties in consolidating and planning regular balanced intermodal shuttles.

While additional cases are required to advance the findings further, the cases in this book elucidate reasons why ports may not be controlling or capturing hinterlands through the strategies of integration that the port regionalisation concept suggests. An argument is also made for greater disaggregation of the factors that challenge or enable port regionalisation processes, including alignment of the institutional models of private sector stakeholders with that of public sector planners and funders. It may be more accurate to state that port regionalisation can only occur as long as a set of favourable commercial and institutional conditions are maintained. While the findings from the cases presented in this book suggest that it is not easy to maintain such conditions, some examples of best practice have shown that they can be altered. For instance, the commercial conditions can be altered (e.g. port terminals taking a direct role in managing hinterland rail services), as can the institutional conditions (e.g. institutional adaptation to allow port authorities to take direct investments in the hinterland). This best practice has been isolated through the understanding of the spatial and institutional characteristics of intermodal transport and logistics as pursued in this book.

Chapter 2
The Geography of Intermodal Transport and Logistics

Introduction

This chapter introduces the key themes of the book, charting the rise of intermodal transport in industry, government and academia, as well as its relation to the wider geography of freight transport such as ports, rail operations and logistics. The academic literature on port and hinterland development is then reviewed to establish the conceptual framework for this research, based on the port regionalisation concept of Notteboom and Rodrigue (2005), which extended earlier spatial models of port development with a focus on institutional relationships governing the complexity of inland connections. The concept accounts for the fact that modern ports operate in an increasingly complex and sophisticated transport and logistics environment, embedded within multi-scalar planning regimes. By using this conceptual framework to analyse a selection of case studies, this research will explicate the spatial and institutional characteristics of intermodal transport and logistics.

The Geography of Intermodal Transport and Logistics

The Origins of Intermodal Transport

Multimodalism is the use of more than one mode in a transport chain (e.g. road and water); intermodalism refers specifically to a transport movement in which the goods remain within the same loading unit. While wooden boxes had been utilised since the early days of rail, it was not until strong metal containers were developed that true intermodalism emerged. The efficiencies and hence cost reductions of eliminating excessive handling by keeping the goods within the same unit were apparent from the first trials of a container vessel by Malcom McLean in 1956.[1] The initial container revolution was thus in ports, as the stevedoring industry was transformed in succeeding decades from a labour-intensive operation to an increasingly automated activity. Vessels once spent weeks in port being unloaded manually by teams of workers; they can now be discharged of thousands of containers in a matter of hours by large cranes, with the boxes being repositioned

---

1 See *The Box* (Levinson, 2006) for a historical account of the advent of containerisation.

in the stacks by automatic guided vehicles. This in turn means that ships can spend a much higher proportion of their time at sea, becoming far more profitable.

As shipping and port operations were transformed by the container revolution, a wave of consolidation and globalisation took place. Shipping lines grew and then merged to form massive companies that spanned the globe. Container ports expanded out of origins as general cargo ports, or were built entirely from scratch. Some existing major ports today show their legacy as river ports and require dredging to keep pace with larger vessels with deep drafts (e.g. Hamburg), whereas newer container ports are built in deep water, requiring not dredging but filling in to create the terminal land area (e.g. Maasvlakte 2, Rotterdam). The move to purpose-built facilities with deeper water severed the link between port and city, with job numbers reduced and those remaining moved far from the local community, altering the economic geography of port cities (Hesse, 2013; Martin, 2013).[2]

Most of the new generation of container ports are operated by one of a handful of globalised port terminal operators such as Hutchison Port Holdings or APM. This is the result of the trend towards consolidation across the industry in the decade leading up to the onset of the global economic crisis in 2008, in which many mergers and acquisitions took place in both shipping liner services and port terminal operations (Slack and Frémont, 2005; Notteboom, 2007; Song and Panayides, 2008; Van de Voorde and Vanelslander, 2009; Notteboom and Rodrigue, 2012). In 2012, the top ten carriers controlled approximately 63 per cent of the world container shipping capacity (Alphaliner, 2012), while the top ten port terminal operators handled approximately 36 per cent of total container throughput (of which 26.5 per cent was just the top four), measured in 'equity TEU' (Drewry, 2012).[3]

With liner conferences[4] creating additional horizontal integration, shipping lines were able to benefit from massive scale economies, further reducing the price of ocean freight. This helped to drive globalisation strategies, which in turn fuelled the container shipping boom of the early twenty-first century. Container movements increased not just in relation to actual trade but from the rise of transhipment due to increasingly complex liner networks involving hub-and-spoke strategies. This means that a loaded shipment may travel much further than it would if it were to go directly between the two ports nearest to the origin and destination. Container handlings at ports also swelled due to the increasing number of empty container

---

2   For more detailed analyses of port-city relationships see Hoyle (2000), Hall (2003), Ducruet and Lee (2006), Lee at al. (2008), Hall and Jacobs (2012), as well as a selection of papers in Wang et al. (2007).

3   The 'equity TEU' concept was devised by Drewry as a more accurate way than simple TEU throughput to account for the fact that some terminal operators have shares in one another.

4   Shipping lines can operate 'alliances', through which they share space on one another's vessels. 'Conferences', where they act as a 'public cartel' and set joint prices, are legal in some countries but were ruled illegal in the EU in 2008. Some countries allow them to reduce destructive competition (e.g. price wars that damage the market).

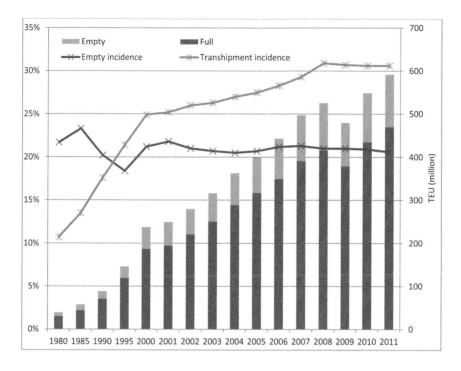

Figure 2.1   Loaded and empty container movements as shares in total world container movements
Source: Author, based on Drewry (2013)

movements resulting from trade imbalances between exporting and importing regions. Figure 2.1 shows total container handlings at world ports, divided into full and empty, as well as the incidence of empty movements and transhipment.

The figure shows that, while the number of empty container handlings has risen sharply, the percentage of total handlings has changed little since 2000. The interesting statistic is the increasing incidence of transhipment, meaning that in 2011 30.6 per cent of container handlings at world ports were not genuine trade but containers being transhipped as part of a hub-and-spoke or similar liner strategy.

As well as horizontal integration through acquisition and merger, there has been much vertical integration, with shipping lines investing in port terminals (e.g. Maersk/APM and others). The increasing integration between shipping lines and ports has created an almost entirely vertically-integrated system from port to shipping line to port within the same company. The inland part of the chain is the new battleground but it is more complex than the sea leg, as will be seen in this book.

As a result of these changing industry dynamics, ports changed from city-based centres of local trade to major hubs for cargo to pass through, with distant origins and destinations. This development was driven to a large degree by the container

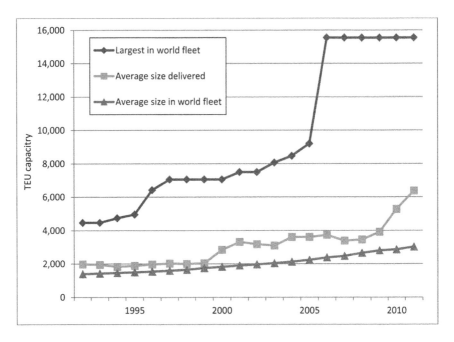

Figure 2.2   Container ship size progression 1992–2011
Source: Author, based on Drewry (2013)

revolution, as distribution centres (DCs) located in key inland locations became key cargo generators and attractors (see later section on distribution). Port hinterlands began to overlap as any port could service the same hinterland. Another key factor was increased economies of scale available from ever-increasing ship size (see Figure 2.2). The figure shows that the average size of newbuild delivered in 2012 almost matches what was the largest container vessel in service only ten years ago.

Shipping services were rationalised, with large vessels traversing major routes between a limited number of hub ports. Cargo was then sent inland or feedered to smaller ports. The introduction of new, larger vessels on mainline routes is initiating a process whereby vessels cascade down to other trades. The result is that feeder vessels are being scrapped at a higher rate than normal and the order book for new builds is at a historic low. According to a study by Clarkson's, in the first two months of 2013, delivered capacity in the sub-4,000 TEU range was four times less than the amount scrapped (Porter, 2013). This will have serious effects for smaller ports that are reliant on feedering but that cannot accommodate ships above 1,000 or 2,000 TEU, leading to new battles for second-tier regional and peripheral ports (Wilmsmeier and Monios, 2013).

The introduction of ever-larger vessels on mainline routes is attractive for shipping lines but will strain ports severely. This was known from the early days of containerisation: 'The ship designer has always been the pacemaker in shipping transport innovations since his creation has merely to float and sail economically per ton mile; whereas the port engineer has to cope not only with the demands of ship designers, but also with the physical difficulties of the port's land and water sites' (Bird, 1963: p.33). Ports invest large sums upgrading their facilities and competing to receive vessel calls, but handling such demand spikes is difficult. Large container drops can result in inefficient crane utilisation, as the numerous large cranes required to service large ships are not all required between calls; furthermore, such numbers of containers cannot always be moved in and out of the port in a smooth manner. Additionally, shipping lines already cannot meet their own schedules; current average reliability across the industry is below 70 per cent. The larger the vessel and the larger the drop of containers at each call, the larger the knock-on effect of such poor reliability on the rest of the container system. Some ports may be able to mitigate this challenge through a satellite terminal system whereby containers are pushed into the hinterland on regular shuttles to large intermodal terminals (Veenstra et al., 2012). In order to manage such a process successfully, several practical and institutional barriers will need to be overcome, as discussed in this book.

Despite such consolidation in shipping and port terminals, the business of maritime transport remains highly volatile, not just cyclical but dramatically so, exhibiting widely divergent peaks and troughs. According to a senior executive from Maersk: '2009 for Maersk Line was the worst year we have ever had and 2010 was the best – that is not very healthy' (Port Strategy, 2011). Therefore, port actors seek stability where possible, needing to anchor or capture traffic to make themselves less susceptible to revenue loss when the market is low. Inland transport is now the area where they seek to secure this advantage. This need to control inland connections is not just about physical infrastructure but institutional issues such as labour relations and other regulatory issues. For example, the labour difficulties in the ports of Los Angeles and Long Beach in recent years have increased the likelihood of Mexican ports competing as gateways for North American trade. But they can only do so if they have high quality infrastructural connections within Mexico and will need to overcome organisation challenges within the Mexican rail industry (Rodrigue and Wilmsmeier, 2013).

While the rise of intermodalism originally related primarily to sea transport, the land leg was undertaken by all modes, which were also busy transporting domestic traffic. The increasing integration of international and domestic transport was a result of globalised supply chains growing out of relaxed tariff and trade barriers as well as ever-cheaper sea transport. Inputs to manufacturers and even finished products were being imported at a growing rate from cheaper supply locations and, to overcome congestion and administrative delays at ports, shipping lines began to offer inland customs clearance. The bill of lading could now specify an inland origin and destination. These sites were variously known as inland clearance depots (ICDs) and 'dry ports'. In the UK, so-called 'container bases' were established at key locations around the country in the late 1960s to handle containerised trade passing through southeastern ports to and from inland locations in the north and centre of the country. These freight facilities were usually located next to intermodal terminals, but the actual transport mode could be road or rail (see Hayuth, 1981; Garnwa et al., 2009). This kind of trade was especially promoted for landlocked countries lacking their own ports. Thus the 'dry port' could offer a gateway role and reduce transport and administrative costs (Beresford and Dubey, 1991).

It was in the United States where true intermodal transport was established successfully. During the 1980s, carriers operating in the transpacific trade were suffering from excess tonnage and low freight rates. To increase its cargo volumes, American President Lines (APL) formed the first transcontinental double-stack rail services, recognising that intermodal transport provided a ten-day time saving compared to the sea route through the Panama Canal to New York (Slack, 1990). While the transit time was important, APL also offered more services to the shipper as the customer could receive a single through bill of lading. The growth of discretionary cargoes allowed APL and other shipping lines to expand their capacity in the transpacific. By using larger, faster ships, a carrier could offer a fixed, weekly sailing schedule, while the additional capacity reduced per-unit costs. In Europe, intermodal freight transport developed in the 1990s, although the fragmented geographical and operational setting (e.g. national jurisdictions and constraints on interoperability) as well as physical constraints (e.g. limited opportunities for double-stack operation) meant that progress was not as swift nor as successful as in the United States (Charlier and Ridolfi, 1994).

Containers and Intermodality

Unitised transport refers to the movement of freight in a standardised loading unit, which may be a single consignment of goods or may be a groupage load of smaller consignments managed by a freight forwarder. The unit in question, often referred to as an 'intermodal transport unit' (ITU) or 'intermodal loading unit' (ILU) may be an ISO maritime container, a swap body or a semi-trailer.

ISO containers are the strongest loading unit, as well as being stackable. They are, therefore, the most versatile. The key underpinning of successful intermodal transport was not simply the invention or adoption of these containers but their increasing standardisation. This was a long process (see Levinson, 2006 for detailed history and the role of the ISO), that resulted in a handful of main container types. While there still remain several lengths, heights and widths, 20/40ft long units remain dominant on deepsea vessels and containers are therefore measured as multiples of 20ft (twenty-foot equivalent units or TEU). Significant divergence remains, however, particularly domestically. For example, both the UK and the USA favour domestic intermodal containers of the same dimensions as articulated road trailers, for obvious reasons (45ft in the UK and 53ft in the USA).[5] Standard

---

5   Deepsea container vessels are fitted with cellular holds with twistlocks at regular intervals to hold multiples of 20ft units and are therefore unable to mix these containers with 45ft or 53ft boxes. Specialist intra-European vessels are fitted for 45ft boxes.

height is 8ft6 although other heights exist, and 9ft6 (known as high-cube) are increasingly common as they allow extra volume, subject to weight limits. Standard width is 8ft, although again other widths are possible, and in Europe the 8ft2 (known as pallet-wide) is popular because, again, it is closer to the load capacity of a semi-trailer.

Swap bodies can be moved between road and rail vehicles, but are not strong enough to be stacked or to be used on sea transport. They can be fully rigid or curtain-sided for side loading. Finally, semi-trailers are the common sight on today's roads, consisting of a loading unit integrated with the trailer.[6] Again, these can be rigid or curtain-sided or whatever formation is suitable for the cargo. ISO containers, swap bodies and semi-trailers can also carry temperature-controlled goods, with their own integrated refrigerating units (requiring a regular power source). Road vehicles can also be carried on rail wagons in their entirety (as in the Channel Tunnel). This is referred to as 'piggyback', and is less common than utilising a container (Lowe, 2005; Woxenius and Bergqvist, 2011).

Despite the impression that intermodal transport is a seamless journey from origin to destination, a massive amount of effort is required to underpin this apparent integration between land and sea transport. From a mobilities perspective, true intermodality can be considered as an attempt to annihilate the difference between land and sea (Steinberg, 2001; Broeze, 2004), to produce what Martin (2013) calls an integrated 'logistics surface'. Yet the different characteristics of land and sea space mean that such visions of seamless intermodal movement obscure a very fragmented reality. Increasing standardisation has been essential to the development of intermodal transport, not only in the physical standards of containers and handling apparatus, but in domestic and international regulation, in business practice and information sharing, and in supply chain integration through mergers and acquisitions.

The institutional approach in this book represents an attempt to tease out some of these structures and actors that can facilitate or constrain such intermodal integration. Looking solely at transport costs risks promoting the view that it is simply a seamless move, whereas a recognition that this move can only take place if several successful structures are established and maintained allows a more nuanced view and a recognition of potential barriers to successful intermodal transport and logistics.

Distribution and Logistics

Globalisation and specialisation have resulted in the spatial separation of production and consumption. Gateways for international cargo interact with national, regional and local hubs to articulate joins between international and domestic flows. These articulation points or nodes (see next section) can have different features; for

---

6   Trailer is the preferred term in Europe for the wheeled unit on which a container or swap body rests, while chassis is used in the United States.

example, they may be transport interchanges or they may be large distribution centres where various supply chain activities take place. The severance between the port and the city as observed earlier was followed by a similar rupture between inland freight handling centres and their city locations. Hesse (2008), drawing on Amin and Thrift (2002), identified new 'geographies of distribution', remarking that 'the freight sector reveals an astonishing degree of disconnection of logistics networks from traditional urban and economic network typologies' (p.29).

Distribution has progressed from a simple transport procedure to an integrated system based on large distribution centres, which have transformed from simple storage warehouses into large buildings with storage, cross-docking, customisation, light processing and information management. They represent a concentration of logistics processes that might previously have required many separate companies and several locations. Up and downstream supply chain integration will be covered elsewhere in the book, but the key point is that, rather than selecting locations close either to production or consumption, these new sites were located at intermediate positions, suiting their new role as centres of distribution rather than centres of production or consumption (see Table 2.1).

Table 2.1   Centralised and decentralised distribution hubs

	Function	**Location**	**Examples**
City	Traditional place of goods exchange (regional distribution)	Historical urban centres	Market places, traditional locations for urban retail, warehouses
Port city (or inland port city)	Traditional place of goods exchange (long-distance distribution)	Shorelines, large inland waterways, intersections of trade routes	Port land area, inland port land area, warehouses
Urban periphery	Spatial anchor of modern distribution networks	Cheap land and workforce, motorway intersections, on edge of urban areas	Industrial DCs and warehouses, big box retailers and shopping malls
Large scale distribution	Decoupling of distribution from urban market place	Cheap land, workforce and motorway access, intermediate for several urban areas	National or regional hubs for global distribution firms

Source: Adapted from Hesse (2008)

The next section will consider centrality and intermediacy more directly, but the key point is that the changing location of distribution centres illustrates how an intermediate location, suitable for distribution to several urban centres, can itself become a central location, exerting through its agglomeration benefits a centripetal pull on logistics facilities. The result is large concentrations of flows in certain hub regions, such as the Midlands in the UK or the Rhine-Ruhr area in Germany.

Centralisation in the context of logistics can thus inhabit several meanings. From one perspective, central means a DC located in a city, as opposed to intermediate, in which there would be one DC in between several cities, serving all of them. In this view, the precise location of the DC in a city is irrelevant, but from a more local view, moving DCs to suburban peripheries would be considered a process of decentralisation (see Table 2.1). Alternatively, the intermediate strategy of DC location between cities can be viewed as a strategy of centralisation. This is for two reasons; first, because inventory is stored centrally in one DC rather than spreading it across many, and second, because in most cases the intermediate location is roughly in the centre of a country or region. Therefore identification of centrality and intermediacy depends to a large degree on the chosen perspective.

The flexibility of road haulage allowed part loads and frequent small deliveries to support increasingly complex supply chains and new trends towards low inventory levels. Intermodal terminals can be used to support low inventory models, via the 'floating stock' concept, meaning that stock both in transit and awaiting transfer at terminal interchange points is monitored in an inventory system linking store, DC, intermodal terminal and gateway port (Dekker et al., 2009; Rodrigue and Notteboom, 2009). Just as new purpose-built port terminals were built away from their former urban locations, so too were these large distribution nodes, with a focus on the optimal distance to reach several major cities within a region or country, and clustering and agglomeration strategies resulted in large logistics platforms with multiple large tenants.

The notion of transport solely as a derived demand has been challenged and reformulated as an integrated demand (Hesse and Rodrigue, 2004; Rodrigue, 2006; Panayides, 2006). The planning of logistics processes influences transport requirements but the former are themselves influenced by the location and quality of transport nodes and corridors. As such, freight flows can be impeded by networks of nodes and hubs that may not perform their key functions adequately. Thus the (current and proposed) functions of these nodes need to be understood properly in order to incorporate their effects into an economic geography of freight transport. Supply chains may be forced into sub-optimal paths which are then exacerbated by issues of path dependency, decreasing the visibility of alternative options. It has been estimated that an international box movement involves around 25 actors (Bichou, 2009), therefore it is a complex process in which many aspects have low visibility.

Transport operators have increasingly rebranded themselves as providers of logistics services. This is more than just marketing but a recognition of how transport requirements are derived from and in turn exert their own influence upon logistics decisions. From an organisational perspective, there is first the issue of who makes the decisions. Often a small shipper will contract a freight forwarder to manage the transport process, including grouping many small consignments together. A larger company may have an in-house logistics division or they may outsource logistics management to a third-party logistics provider (3PL).[7]

The redefinition of transport as an integrated demand relates to the fact that the dominance of 3PLs integrates the supply chain to the extent that transport demand is not simply derived from the independent decisions regarding location of production and consumption, but is part of a unified strategy linking all processes. This represents the apogee of a process of increasing integration over the second half of the twentieth century, where large globalised 3PLs now manage the entire movement of goods within a global supply chain (Hesse, 2008). One of the aims of this book is to explore these processes of integration, revealing where the key sites of interaction between such processes can be identified, and whether increasing integration of global supply chains results in controllable transport corridors in the same way as has been observed from the integration in maritime transport. This is of particular relevance because intermodal transport must be analysed more accurately as intermodal logistics; making the trunk haul feasible by rail requires a suitable distribution strategy based on several factors and processes, such as appropriately located distribution centres, integrated planning and the characteristics of order types and sizes. These issues will be explored in the case study in Chapter 5.

Nodes, Networks and Corridors

A node may be defined simply as a location or a point in space; in the case of transport this would represent an origin or destination of a linkage. In practice, only nodes of a certain size are relevant, where a certain level of units are concentrated, move through or otherwise utilise this access point. A node can serve as an access point to join a transport network or it may be a point joining two linkages within a system. Two defining characteristics of such nodes are centrality and intermediacy (Fleming and Hayuth, 1994). Centrality can be derived from location theory (Von Thünen, 1826; Weber, 1909; Christaller, 1933; Hotelling, 1929; Lösch, 1940), in which the centre is the marketplace and location of important administrative and government activities, exhibiting a centripetal pull on the region, while intermediacy refers to an intermediate location in between such centres. From a transport perspective, Fleming and Hayuth (1994) observed how central locations are often also intermediate, acting as gateways to other locations, They added that such locations can be manufactured, depending not solely on natural geographical endowments, but on commercial or administrative decisions (see also Swyngedouw, 1992). Ng and Gujar (2009) discussed centrality and intermediacy as determining concepts of inland nodes and how they can be affected by government policy.

Nodes can also be defined as points of articulation which are interfaces between spatial systems (Rodrigue, 2004), particularly different levels (e.g. local

---

7 Also referred to as logistics service providers (LSPs).

and regional) and types (e.g. intermodal connections), but the articulation concept can also include joining different categories of system, in this case transport and logistics systems. This involves the relation of the transport activity to other related activities such as processing and distribution, all activities within the wider logistics system (Hesse and Rodrigue, 2004). This role of the node as an articulation point between transport and logistics systems will be a recurring theme during this research and forms the basis of the requirement for an institutional approach.

Locations, points or nodes are joined by links. These links may firstly be physical, meaning either fixed, such as roads, rail track and canals, or flexible links such as sea routes. They may also be operational links, referring to services such as road haulage or shipping schedules. Links can be rated in terms of their capacity, current usage, congestion and other operational categories. Nodes are often measured by their connectivity, which again could either refer to the number and quality of physical links or the number and frequency of operational links, all of which derive to a certain extent from the qualities of centrality and intermediacy already discussed. Operational strategies of freight operators go beyond single links and can be expressed in various ways such as hub-and-spoke, string or point-to-point. These combined operational plans then become transport networks, either a single company network or the accumulation of all available services within a given area.

A network can be defined as the set of links between nodes. Again, this may be considered from a physical or operational perspective. The issue of connectivity just discussed can be used to assess the quality of a node but can also refer to the quality of a network, in which a number of nodes are connected. A high quality network may contain a number of nodes with high connectivity, high centrality and high intermediacy, linked to each other with frequent, high capacity services within a small number of degrees.

A corridor can be defined as an accumulation of flows and infrastructure (Rodrigue, 2004). In some ways the concept of a corridor is somewhat arbitrary and may be used for branding or PR purposes. This is because, beyond a specific piece of infrastructure (e.g. one road or rail line between two places), a corridor usually denotes a large swathe of land through which multiple routes are possible along numerous separate pieces of infrastructure with many different flows organised and executed by different actors. The corridor branding concept can be useful for attracting funding and focusing attention on a specific region, for example connecting a port with an inland area (see the Heartland Corridor example in Chapter 6).

Scholars often prefer a network focus to a corridor focus because it gives a better measure of the wider distribution requirements of each node, whereas a corridor focus can be limiting, given the globalisation of today's distribution patterns. Yet a corridor approach is more amenable to public planners, who need to coordinate many divergent demands for transport and land use, within local, regional, national and international policy, planning and funding regimes.

Intermodal Terminals, Logistics Platforms and Corridors

A transport terminal is an interchange site, a node on a transport network. An intermodal terminal is a site where mode is changed, generally road/rail or road/ barge. These can be as simple as a rail siding (basically just a spur of rail track off the main line) with a small area for a mobile crane or reach stacker to lift the cargo, or it may be a large area with several tracks and large gantry cranes. Figure 2.3 shows a small intermodal terminal at Coslada, near Madrid in Spain. It has two marshalling tracks and four handling tracks, one gantry crane and three reach stackers for container handling, and some space for stacking containers alongside the tracks. The triangular area is for empty storage and there is an administration building and a customs building. The contrast is clear when observing a much larger site at Memphis, USA (Figure 2.4 below).

This site has eight gantry cranes, five for operating the handling tracks and three for the stack. The site has 48,000 feet of handling track with enough length to work a full train of 7,400ft without cutting. The terminal has a capacity of over 500,000 containers per year. A large marshalling area can also be seen to the right of the terminal itself.

The size of an intermodal terminal will depend on its role and how many functions it provides. Trains coming from the mainline will often need to be marshalled in yards beyond the perimeter of the terminal itself. They may need to be split into sections for different parts of the terminal or simply because many terminals are not long enough to handle a full length train. This is especially the case in the United States with very long trains reaching over 10,000ft in some cases (meaning that, with double-stacking, US trains can reach capacities of 650 TEU, compared to around 80–90 TEU in Europe). Additional staff and shunting locomotives are required for this purpose, before the wagons are in place for unloading and loading to commence. Then the train sections will be brought into the site and onto the handling tracks, some of which will be under cranes and others which will be just for marshalling or storage.

The handling can be done by reach stackers or by fixed gantry cranes, depending on the size and layout. They can be grounded or wheeled, in the American terminology. In Europe, intermodal terminals are generally grounded facilities, meaning that containers are transferred between train and truck, and if a direct transhipment is not made, the containers are stacked on the ground. The truck driver will arrive at the terminal with a trailer or chassis and the container will be lifted onto this. By contrast, in the US, both grounded and wheeled facilities are found. This is because trailers and containers are owned by the carrier (be that the shipping line or 3PL), so the truck driver simply arrives at a wheeled site in a tractor. Containers are loaded onto waiting trailers and the arriving driver will hook up to a loaded trailer and take it away. These wheeled facilities require a great deal more room as there is less equipment that can be stacked, but they can be quicker for the incoming drivers who do not have to wait for their container to be located in a stack. This also means that cranes make fewer unproductive moves to pick

Figure 2.3   Small intermodal terminal at Coslada, Spain
Source: Imagery: DigitalGlobe. Map data: Google, basado en BCN IGN
España

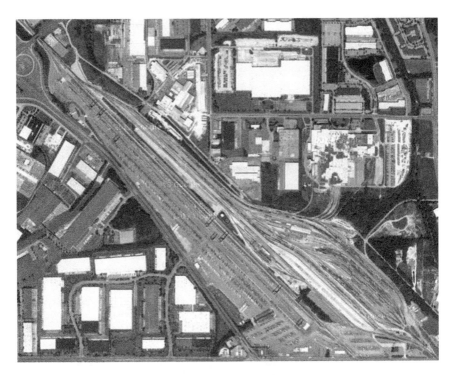

Figure 2.4   Large intermodal terminal at Memphis, USA
Source: Imagery: DigitalGlobe, State of Arkansas, USDA Farm Service
Agency. Map data: Google

through a stack of containers. They are also less capital intensive than grounded facilities because they require less specialised handling equipment (Talley, 2009).

An intermodal terminal can be operated by a transport provider as part of their transport network or it can be operated by a dedicated terminal operator handling trains from multiple individual rail operators. A terminal requires a small office building, and will often provide some basic services such as container cleaning and maintenance and some space for an empty depot. If the cargo is international, a secure building will be required for customs inspection (see earlier discussion of ICDs and dry ports).

The basic function of the terminal is to change mode, thus it can be a site where many trucks bring or collect containers to and from the rail head. By contrast, intermediate supply chain activities can be performed there. It could be very basic, such as some container stripping and reloading, or combining small loads into groupage loads, or LCL (less than container load) into FCL (full container load). These operations are performed in a container freight station (CFS). This would normally be the limit of what would be provided in even a large intermodal terminal. Beyond that, there will either be individual organisations or 3PLs with

Figure 2.5   Magna Park logistics platform in the UK
Source: Imagery: DigitalGlobe, Getmapping plc, Infoterra Ltd, Bluesky.
Map data: Google

their warehouse or distribution centre located nearby, or they could be grouped together in a large logistics platform, which is a multi-user site with shared facilities (see example in Figure 2.5).

A multi-user logistics platform will contain many large buildings and will have its own governance structure, in terms of development, ownership and operation, including selling and leasing of plots and provision of shared services such as cleaning, security, post, catering and so on. The link between the logistics platform and the intermodal terminal is that some platforms will have an intermodal terminal as part of the site (see example of Bologna in Figure 2.6 below), or may have one located nearby, or may simply have road access only. This relation between the terminal and the logistics activities (whether individual companies or as part of an

Figure 2.6   Freight village with two intermodal terminals at Bologna, Italy
Source: Imagery: Cnes/Spot Image, DigitalGlobe. Map data: Google

integrated site) will be addressed in this research, showing how integrated supply chains relate to successful intermodal transport, through the need to plan loads to keep trains full on regular schedules. Relevant issues such as the roles of the public and private sectors in developing such sites and different operational models will be examined in the empirical research in this book.

For an intermodal terminal to be successful, regular traffic is required (Bergqvist et al., 2010), which generally means a large amount of production or consumption nearby with a suitable distance to origin or destination to support regular long-distance trunk hauls where rail or barge is the natural mode. Various break-even distances have been suggested in the literature (usually averaging at around 500 km), but the reality is that it depends on operational considerations. The longer the distance, the more likely that the increased handling costs of changing mode from road to rail/barge will be offset by the cheaper per-unit transport cost. However, this depends on the quality and capacity of the intermodal infrastructure as well as suitably scheduled services at the right departure and arrival times, without unnecessary delays along the route. It also depends on the total quantity of cargo, as such services will not be economic unless they achieve high utilisation in both directions. For these and other reasons, road haulage still retains a large proportion of medium and even long-distance flows. At short distances, road obviously has the advantage in most cases, but it has proven possible to run intermodal services at short distances, if very high volume is achieved, with good timetables allowing quick turnaround and high utilisation of expensive rail assets. Finally, the administration savings from avoiding port congestion can be another reason to choose an intermodal shuttle, which may offset the higher transport cost. That is why any intermodal scheme (terminal or corridor) must have a clear business model, relating both to transport cost savings (assessing the base transport cost as well as loading and capacity utilisation considerations) and logistics cost savings (including assessment of administration, customs clearance, storage and delays).

For planners to support such intermodal corridors, funding for specific infrastructure must be aligned with economic imperatives of the regions through which the corridor passes, requiring governance initiatives or simply branding to focus attention on the various links and flows of which the corridor is comprised. One example is the high capacity 20-mile Alameda Corridor enabling the ports of Los Angeles and Long Beach to bypass congested areas around the port to access their hinterlands (Jacobs, 2007; Rodrigue and Notteboom, 2009; Monios and Lambert, 2013b). This project was pursued by the public port authorities with funding support from the federal government and other agencies (see Chapter 6). The Betuweroute is a similarly high capacity line (double tracked and with double stack capacity) linking the port of Rotterdam with the German border, a distance of 99 miles. The Dutch section was opened in 2007, but the German section has not yet been completed. The existing section in the Netherlands was built and funded by the Dutch government with support from the European Union through the TEN-T programme (van Ierland et al., 2000; Lowe, 2005). Both of these corridors have the aim of enabling large congested ports to move containers in and out quickly. By contrast, long-distance corridors in Africa are more focused on providing access to global markets for locations deep in the interior, especially landlocked countries. The Central and Northern Corridors link the ports of Dar es Salaam, Tanzania and Mombasa, Kenya with their respective national hinterlands as well as to landlocked countries Uganda, Rwanda and Burundi. Covering such vast distances, these corridors represent actual infrastructure as well as a variety of operators and interests, resulting in governance structures that attempt to harmonise these interests and attract investment to resolve infrastructural and operational difficulties (Adzigbey et al., 2007; Kunaka, 2013). Lowering transport costs on these corridors is not simply a matter of upgrading infrastructure but dealing with operational issues such as trade imbalances and equipment shortages, harmonising customs regulations and border crossing formalities, as well as dealing with delays caused by congested handling facilities in the ports. Intermodal corridors, like terminals, can therefore be based on straightforward transport priorities achieved through the provision of high capacity infrastructure or they can be related to the institutional difficulties of resolving administrative and logistical issues.

The key practical issues in rail operations relate primarily to planning difficulties. Once a flow is identified, rail operators need to timetable the service including driver hours and changes, intermodal terminal interchanges and so on, then book a path on the network and pay the track access charge, as well as purchase or hire locomotives and wagons. Wagon sizes need to be matched with container sizes. For instance, in the UK international flows will be in 20ft and 40ft containers, which are generally carried on 60ft wagons. Yet domestic containers and short sea containers are 45ft long, thus requiring a shorter wagon (usually 54ft, therefore still wasting capacity), that will conflict with flows of maritime boxes. Rail operators generally prefer shuttles of fixed wagon sets to reduce such difficulties, which may result in wasted capacity at times. Woodburn (2011) surveyed load factors on UK freight trains and found that, on average, only 72 per cent of wagon capacity was filled on existing services.

From a network perspective, the last two decades have seen a decline in wagonload service (where operators would pick up loads from a number of individual, private sidings) to a majority of full trains, between load centres, usually aggregated loads on behalf of shipping lines or 3PLs. This relates to the rise of large intermodal terminals and dedicated port shuttles, which are one of the key aspects examined in this book.

Whereas weight and speed restrictions limit the capacity of road haulage (e.g. maximum of 44 tonnes and 56 mph in the UK), the natural advantage of rail is that trains can reach high speeds and make good time across long stretches of straight track. In practice, the average speed in congested areas such as Europe is quite slow, due to the many delays, bottlenecks and time spent in sidings waiting for passenger trains to pass. On the crowded European network, passenger trains are generally given priority. This is less of an issue in other parts of the world such as the United States where passenger rail is in the minority and the tracks are owned by the private freight companies.

Government Interest in Intermodal Transport

A key challenge of transport geography is to understand shifting notions of infrastructure provision brought about by the changing roles of the public and private sectors (Hall et al., 2006; Hesse, 2008), while acknowledging the difficulty of predicting the effect of government investment (Rodrigue, 2006).

It is a public responsibility to ensure sufficient capacity on all transport links to support a growing economy, but the mix of public and private interests in freight operations can result in considerable uncertainty when it comes to investment in upgrades and capacity enhancements, or connecting freight nodes to the transport network. While highways and motorways are generally maintained by governments for both passenger and freight use, rail and waterways can be either privately or publicly owned. Interchange sites such as ports and rail/barge terminals may exhibit a variety of ownership, management and governance regimes (as discussed in this book). Where they are under public control, rail and waterways are for the most part simply maintained in their current state, with the occasional new section or upgrade, but the high levels of public investment expended for the apparent benefit of private companies can be contentious. The success of intermodalism in the United States is partly a result of a vertically integrated system in which rail operators own and manage both their infrastructure network and the operations. The US is large enough to sustain competition between different operators each with their own extensive infrastructure network serving most of the same origins and destinations. A smaller geographical region such as Europe would find such a system difficult; therefore, while this was the original model when rail was first developed in the UK, it was eventually unified under nationalisation (before being reprivatised under a different model – see Chapter 5). The current system across Europe is that the infrastructure is owned by national governments while individual rail operators compete with one another to run services, paying access charges for their use of the track infrastructure. In the UK these companies are private, whereas in Europe they are a mixture of private and public. However, even where they are public, evolving rail reform in Europe due to EU policy has required that they operate as quasi-private companies, with full organisational separation from the infrastructure-owning parts of the same national companies. This was supposed to increase competition with benefits for the user, but there are different views on whether this has simply increased fragmentation. An interesting comparison is China, which is still publicly controlled, divided into several vertically integrated regions. System reforms (in particular ways of introducing competition) have been mooted over the years (e.g. Wu and Nash, 2000; Xie et al., 2002; Pittman, 2004; Rong and Bouf, 2005); recently Pittmann (2011) suggested that in a country with large distances and large volumes (comparable to the United States), parallel competition could be introduced in China through multiple closed systems.

From an operational perspective, in terms of the impact of rail infrastructure on successful intermodal operations, there are rail gauge (width between the rails) compatibility issues between some countries, such as between continental Europe and the Iberian peninsula, and in the UK loading gauge (width and height) is restricted due to bridges and tunnels, meaning that high cube containers cannot be carried on some parts of the network unless low wagons are used, adding expense and inconvenience. The other site of interaction between infrastructure and operation is intermodal terminal development (discussed in Chapter 4).

In Europe, most rail networks were managed by the national government until recent times (Martí-Henneberg, 2013), and terminals were developed both by private transport operators attached to the national network and by the national rail operators themselves. These sites are now mostly owned and/or operated by private operators (e.g. UK examples discussed by Monios and Wilmsmeier, 2012b), or, in a liberalised EU environment, the vertically-separated and quasi-private but still nationally-owned rail operator (e.g. European examples discussed by Monios and Wilmsmeier, 2012a). In other countries, the rail operations remain wholly or predominantly under state control (e.g. India – see Ng and Gujar, 2009a, b; Gangwar et al., 2012). In the United States, where rail is privately owned and operated on a model of vertical integration, intermodal terminals are developed and operated by the private rail companies (Rodrigue et al., 2010; Rodrigue and Notteboom, 2010).

Until recent times, operational decisions and mode choice were the preserve of the industry. The inland leg was taken by road, rail or inland waterway, according to the economic and practical imperatives of the shipper and transport provider. Rail and water generally dominated long hauls because they were cheaper, whereas road haulage would perform shorter journeys, particularly any journey where its natural flexibility and responsiveness made it the natural choice.

The role of the public sector was operational in some countries where rail was nationally owned, but otherwise related mostly to regulation (see discussion of the influence of the Staggers Act in the US in Chapter 6). However, as emissions and congestion rose up the government agenda in the 1990s, governments began to see their role as more directly interventionist in order to address the negative externalities of transport. The European Union transport policy document published in 2001 (European Commission, 2001) made a clear statement in favour of supporting intermodal transport as one method of reducing emissions and congestion (see Lowe, 2005 for more detailed discussion of intermodal policy development in the EU).

In the late 1990s and early 2000s, policy documents proliferated across Europe promising support for greener transport measures to reduce dependence on road transport, while also taking care politically not to be seen to threaten the performance of the road haulage industry which remains essential to a functioning transport system. While road haulage produces more emissions than other modes per tonne kilometre, lorries are increasingly environmentally friendly, due in part to government regulation. For example, in the EU, successive engine regulations have proceeded from Euro I in 1992 up to Euro VI in 2013, each one progressively reducing the amount of carbon, NOx and particulate matter permitted per kWh. Factors encouraging modal shift away from road haulage include continuing fuel price rises and, in Europe, the working time directive limiting driver hours per week (although there are questions as to how closely such regulations are followed in different parts of Europe, not to mention the lack of such regulation in other parts of the world). Road user charging is another policy implemented in some parts of Europe (e.g. the maut scheme implemented in Germany which charges lorries for use of the motorways). Better fleet management, use of ICT, increased backhauling, triangulation, reverse logistics, returning packaging for recycling and other operational measures (McKinnon, 2010; McKinnon and Edwards, 2012), mean that emissions (if not congestion) can be reduced quite substantially through improvements to road operations rather than through modal shift.

Road haulage remains the natural choice for short hauls due to its flexibility and convenience. Rail and water modes were already being used where they were cheaper or faster, i.e. long hauls. The policy aim around the turn of the century was now to encourage the use of rail and water on medium-length hauls, and to promote whatever actions could enable this, such as harmonising regulations, improving transport infrastructure and so on. Visibility is also key, as there remains a feeling that road haulage is often selected by shippers and forwarders out of habit and a lack of knowledge, experience and familiarity with alternative modes, exacerbated by a fear that they are more difficult or unreliable (RHA, 2007). The result is various policy goals and instruments designed to stimulate intermodal transport.

The establishment of the Single European Market in 1993 and the increasing integration throughout the EU, including customs union and almost complete currency union, altered distribution strategies and increased cross-border freight movements, including a change in the location of DCs as companies developed pan-European rather than national distribution strategies. A series of European directives drove progress towards harmonising administrative and infrastructure interoperability between member states.

A cornerstone of these efforts in Europe is the TEN-T programme, which identified high priority transport linkages across Europe; member states can then bid for funding to invest in upgrading these links. It covers both passenger and freight and includes all modes (as well as 'motorways of the sea'). Its primary goal is not modal shift per se but increased connectivity between member states.

Another government goal is to promote economic development to stimulate employment and industry in underperforming regions. This incentive is generally promoted more heavily at local and regional level, even if the funding is often coming from a national or supranational (e.g. EU) source. Economic geographers have debated whether this is a zero sum game (i.e. whichever region gets the new business, the benefit for the country is the same), and this potential conflict of policy goal between intermodal terminals and infrastructure to reduce emissions vs logistics platform development for job creation requires further research (Monios, 2014).

Logistics clusters have many agglomeration benefits both for business and for transport. However, while access to a large transport corridor, especially an intermodal corridor, may reduce emissions over the length of the journey, it will increase emissions around the access point where traffic is congregated. Linking a town to a nearby corridor may bring economic benefits through direct and indirect jobs, but it may increase emissions and raise property prices and other aspects explored by economic geographers. Increasingly, transport geography looks to economic geography to investigate how transport policy links with economic development policy. Much European funding for transport projects is aimed at reducing emissions but is actually pursued by local and regional bodies because they desire economic benefits from logistics development (Monios, 2014).

The role of the federal government in the USA is examined in Chapter 6. Its transport role has been primarily related to safety and licensing regulation, but it is increasingly taking a direct role in intermodal infrastructure and operations, aiming to promote both emissions reduction through modal shift and economic growth through improving access for peripheral regions. For example, the TIGER programme as part of the US stimulus package provided $1.5bn in federal funding in 2009, to be bid for by consortia of public and private partners across the country (see Chapter 6 for full discussion).

These incentives are less common in developing countries, which are focused more on developing their logistics infrastructure to support business (see Chapter 8 for international comparison), therefore interventionist transport policy has been pursued primarily in developed countries. However, supranational development agencies such as UNCTAD and UNESCAP have promoted policy actions to improve port-hinterland connections and logistics performance in developing countries, especially for landlocked or otherwise poorly-connected inland regions (e.g. UNCTAD, 2004; UNESCAP, 2006; UNESCAP, 2008; UNCTAD, 2013). One of the aims of this book is to elucidate the development process in order to improve evaluation of whether government money is being used effectively to achieve its aims. For example, in many cases there are good business reasons why intermodal transport is not flourishing at a certain location due to cargo and route characteristics. Government money is not always the answer unless the industrial organisation can be improved, thus requiring an appreciation of the institutional challenges to intermodal transport and logistics.

The other important role for governments is the regulation, administration and bureaucracy of trade facilitation. Within a country, it will involve licences to operate transport vehicles, regulation of vehicle and infrastructure quality and quantity and permission to operate as a commercial transport provider. It will also cover planning permission to build a logistics platform, provide connections to electricity and water services, incorporating related issues such as noise for local residents and all the small issues of local planning.

At a larger level, there is all the bureaucracy of international trade. This can include bills of lading and various transport and insurance contracts that must be legally approved, as well as customs legislation in each country, especially when a trade route crosses international borders. In developing countries, much effort is invested by international organisations such as the United Nations and the World Bank to decrease border delays caused by mismatch in customs procedures, physical inspection requirements and information sharing (Stone, 2001; de Wulf and Sokol, 2005; Arvis et al., 2007). Countries are strongly encouraged to adopt paperless customs procedures through such online platforms as ASYCUDA. Trade facilitation measures are considered to be even more important than infrastructure in lowering transport costs in many instances (Djankov et al., 2005).

Summary

The key issues relating to successful intermodal transport and logistics can be derived from the above discussion. On the one hand, cost savings from reductions in handling, from harmonisation of standards in unit loads, double-stacked trains where possible, long-distance full loads to harness the natural benefits of rail. On the other hand, difficulties arise from the lead time for service development, planning and forecasting, backhauls and flow matching, companies needing to work together to share information, consolidate and combine loads and enable regular scheduling to provide a true intermodal service to the customer. Overlaying all of this is the national and international law and regulation of transport and trade, such as customs clearance, bills of lading and harmonising of legal documentation for border crossing and so on.

If policy makers want to understand and promote intermodal transport, they must encompass the whole system from the port to the inland node. This understanding should incorporate processes of centralisation and decentralisation occurring in city regions following land use and labour availability, concentration and deconcentration of international and domestic flows and consolidation and integration of global transport and logistics operators. The following section will translate these findings into a conceptual framework encompassing port hinterland integration in which the empirical cases will then be embedded.

If one single observation from the preceding analysis can encapsulate these issues, it is a comparison of integration processes in land and sea transport. In the past decade, shipping lines and port terminal operators have consolidated and integrated their portfolios through mergers and acquisitions, resulting in a small number of dominant firms. These firms have since benefited from significant economies of scale and scope, enabling them to provide something of a seamless intermodal transport movement from port to port. Many actors are endeavouring to pursue this same trend inland, but for true intermodality to be successful and economically feasible, land transport operations require a similar level of consolidation to that of the sea leg. The different characteristics of land transport, however, challenge this goal. As Graham (1998: p.135) wrote: 'the land-side is characterized by relatively low investment, high operating expenses, little scale incentive to collective operation and a considerable level of unremunerated activity requiring cross payment out of sea freight'. This book investigates the nature of these processes in order to identify the institutional barriers to intermodal transport and logistics.

The next section of this chapter commences the conceptual framework for this book. It examines models of the spatial and institutional development of ports and hinterlands, from which analysis it will develop the key issues to be addressed in the empirical chapters.

Port Development: From Expansion to Regionalisation[8]

Earlier Models of the Spatial Development of Ports

A number of authors have attempted to explain the complex process of port development by proposing different conceptual frameworks. One early influential model was the 'main street' concept outlined by Taaffe et al. (1963), whereby 'since certain centres will grow at the expense of the others, the result will be a set of high-priority linkages among the largest' (p.505). However, the location and functions of the nodes connected along these priority corridors are changing. Whereas in the past these corridors were more static, due primarily to the geographical entry barrier represented by port location, they have become increasingly dynamic.

The 'Anyport' model of Bird (1963) was an early attempt to categorise port development, and his model is still widely referenced today. His model was developed through a study of British ports, and, although this work was written before the advent of containerisation, his model remains useful. Bird also recognised that different parts of the port may be at different stages of development, which means that potentially sub-optimal facilities will still be in use in parts of the port: 'Because of the great capital cost of port installations, it is often cheaper progressively to downgrade a dock or quay in traffic importance rather than scrap it altogether' (Bird, 1963: p.34). The two general development strategies charted by Bird are expansion away from the original town site towards large purpose-built berths with deeper water, and the move towards specialised handling facilities, for example oil products or containers.

Rimmer (1967) discussed the models of both Taaffe et al. (1963) and Bird (1963), producing a five-stage model, while Hoyle (1968) presented a modified version of Bird's model, demonstrating the different stages of development for East African ports that were built in the twentieth century, unlike those in Bird's model that grew out of medieval estuary port sites.

Bird (1971) commented that his model was not intended to fit every port, and he conceded that limitations may be apparent due to the fact that port development models can be based on different factors. While his 'Anyport' model is based on port installations, Bird noted that these structures generally reflect wider issues such as changing ship requirements or developments in hinterland access.

Hayuth (1981) developed the concept of dominant ports or load centres that increase their inland penetration and hinterland capture, very much like in the models of Taaffe et al. (1963) and Rimmer (1967). Hayuth noted that 'it is difficult to weigh the importance of each factor in the development of a load centre port, but a large-scale local market, high accessibility to inland markets, advantageous site and location, early adoption of the new system, and aggressiveness of port management are major factors to consider' (p.160).

---

8   This literature analysis draws on work by the author published in Monios and Wilmsmeier (2012a) and Monios and Wilmsmeier (2013).

Barke (1986) produced a similar model, with an additional focus on decentralisation, whereby some port activities are moved from the port to less congested areas. However, in contrast to some of the inland terminal concepts discussed in Chapter 4, Barke specifically noted that these activities remain 'within the city region, and transport and communications technology ensure that they are within easy contact of the core' (p.122).

Van Klink (1998) suggested port city, port area and port region as summaries of previous port models, and identified the rise of port networks as a fourth stage in port development, including logistical control of inland access as a new role for the port in this phase of development, particularly related to the integration of activities at non-contiguous sites. The idea is then developed into a discussion of the kind of networks possible, based on directions of interdependence. This development leads to the importance of cooperation strategies throughout the hinterland area, rather than the old model of investment in the port area alone. The concept of selectiveness of core activities points to a strategy of moving non-core activities to other locations, allowing the port to focus on the core activity of container throughput, for example the use of ECT's inland terminals for Rotterdam traffic that will be discussed in Chapter 4. Kuipers (2002) supports this observation.

Later authors have suggested that simplistic models such as Bird's or the UNCTAD generational model (UNCTAD, 1992) are unable to capture the complexity of port infrastructure, operations and services (Beresford et al., 2004; Bichou and Gray, 2005; Sanchez and Wilmsmeier, 2010). Beresford et al. (2004) developed the WORKPORT model as a response to the need to conceptualise the complexity of this operational environment.

Port System Evolution

While early models aimed to classify and explain the spatial aspects of port development, the literature over the last two decades has addressed many other issues that have an impact on port development and should therefore be considered here.

While the development of load centres and priority corridors as discussed earlier results in a handful of large hub ports, it has been predicted that this concentration will eventually reach its limits and invert (Barke, 1986; Hayuth, 1981), leading to a process of deconcentration, a phenomenon discussed recently by various authors (Slack and Wang, 2002; Notteboom, 2005; Frémont and Soppé, 2007; Wilmsmeier and Monios, 2013). This challenge becomes relevant as the port system moves towards concentration, particularly for unitised cargo, leading to significant challenges to hinterland infrastructure. Ducruet et al. (2009: p.359) argued that 'concentration stems from the path-dependency of large agglomerations', while drivers of deconcentration include 'new port development, carrier selection, global operation strategies, governmental policies, congestion, and lack of space at main load centres'. However, existing theory falls short of differentiating between deconcentration that emerges upon failure of a system in a reactive manner and deconcentration that materialises from proactive port development strategies. Thus the drivers of deconcentration processes can be related not only to the port system, but also to transport and logistics networks.

From the perspectives of shipping lines and port operators, cargo concentration has the clear benefit of achieving economies of scale and density (Cullinane and Khanna, 1999). These strategies, in combination with the exploitation of geographically favourable locations, can lead to a level of port concentration that has severe repercussions on the port hinterland, if they are not counteracted by proactive policies and public sector strategies. Notteboom (1997: p.115) argued that the 'future development of the European container port system will primarily be influenced by the technological and organizational evolutions in the triptych foreland–port–hinterland and the outcomes of some current (trans)port policy', and this approach was developed further in the regionalisation concept (Notteboom and Rodrigue, 2005). However, this analysis did not incorporate the effect of private sector strategies as currently experienced in various forms e.g. inland terminal development or port-centric logistics strategies (Wilmsmeier et al., 2011; Monios and Wilmsmeier, 2012b).

The changing role of the port in the transport chain and the greater focus on the terminal rather than the port have become key issues over recent years (e.g. Slack, 1993; Notteboom and Winkelmans, 2001; Robinson, 2002; Slack and Wang, 2002; Slack, 2007). Olivier and Slack (2006) proposed a renewed focus on questions of agency, upon the ability of ports to 'steer their own future' (p.1414). Likewise, Heaver at al. (2000: p.373) asked: 'Will port authorities become fully-fledged partners in the logistics chain, will their involvement be restricted to a supporting role (safety, land-use and concession policy), or might they disappear from the scene entirely? In order to find answers to these questions, further research, in particular more disaggregated empirical research, is urgently required'. Notteboom and Rodrigue (2009) contended that port authorities have been concerned about losing influence to inland terminals, while recognising that there are many benefits to cooperation.

Based on the product lifecycle theory and following Schaetzl (1996), Cullinane and Wilmsmeier (2011) argued for 'location splitting' (standortspaltung) as a means to extend the port lifecycle when limitations in feasible rationalisation, investment and access are reached. Such creation of a subsidiary in the hinterland provides a potential solution that avoids an inevitable decline, caused either by the emerging inappropriateness of the actual port location (e.g. once-central urban ports) or an increasingly competitive environment. One question that arises is whether location splitting as proposed by these authors can be induced by landside-driven factors as well.

The concept of centrality that explains to some extent the formation of gateways can be augmented by the concept of intermediacy (Fleming and Hayuth, 1994), where a large direct hinterland market is not a necessary condition for concentrating large traffic volumes. Instead, discontinuous hinterlands are supported by logistics zones and inland distribution centres that are connected to ports by high-volume transport corridors. In this sense the adjusted definition of hinterland obtains, one that considers core, congruent and extended hinterlands (Sanchez and Wilmsmeier, 2010). In the UK, port-centric logistics is being used as a way for regional ports to compete with mainports (Mangan et al., 2008; Pettit and Beresford, 2009; Monios and Wilmsmeier, 2012b).

Governments can attempt to direct infrastructure development through policies and funding mechanisms to meet objectives of modal shift or economic development. Such approaches are prevalent in Europe (Tsamboulas et al., 2007; Proost et al., 2011; Bontekoning et al., 2004), but government involvement is also becoming more common in large intermodal schemes in the United States. In light of the complex institutional arrangements governing modern ports, the interplay of the private sector and government at different scales is often unclear (Bichou and Gray, 2005).

Hayuth (2007) observed the increasing vertical integration of shipping lines and noted that one result of this behaviour is that port choice is increasingly being determined by landside factors such as intermodal infrastructure. This issue is related to the degree of logistics integration in the supply chain (e.g. Heaver et al., 2000; Heaver et al., 2001; Frémont and Soppé, 2007; Hayuth, 2007; Olivier and Slack, 2006; Notteboom, 2008). A variety of coordination mechanisms are arising to manage this process, such as vertical integration, partnerships, collective action and changing the incentive structure of contracts (Van der Horst and De Langen, 2008). The terminal rather than the port has increasingly become the primary focus of study (Slack, 2007); subsequently the land-side activities of the seaport have come under closer scrutiny (Bichou and Gray, 2004; Parola and Sciomachen, 2009), leading to the inevitable focus on inland terminals.

In this literature a dual trend may be observed: using inland terminals to enlarge the hinterland of the seaport (van Klink and van den Berg, 1998) and the integration of logistics services within the transport chain. This is because inland costs (both transport and value-added services) have increased in relative importance as components of the door-to-door cost (Notteboom and Winkelmans, 2001). Notteboom and Rodrigue (2005: p.302) noted that 'the portion of inland costs in the total costs of container shipping would range from 40% to 80%. Many shipping lines therefore consider inland logistics as the most vital area still left to cut costs'. Increasingly relevant is the recognition that the port's position has changed from a monopoly to a dynamic interlinkage and a subsystem in the logistics chain (Robinson, 2002; Woo et al., 2011a). However, as noted above, Graham (1998: p.135) wrote that 'the land-side is characterized by relatively low investment, high operating expenses, little scale incentive to collective operation and a considerable level of unremunerated activity requiring cross payment out of sea freight', and it remains unclear if this situation has changed.

Notteboom and Winkelmans (2001) discussed the importance of ports and inland terminals cooperating, but expressed doubts over public authorities being proactive in this role. They posited that this problem may be due to the focus of the (public) ports being on expanding the local economy, rather than possessing the nature of a private sector organisation that would pursue the interests of the port itself. More recent strategies of privatisation and corporatisation (see Chapter 8) represent attempts to overcome this problem, while being limited to a handful of major ports.

The extension of a port's influence into the hinterland is one opportunity for port authorities to intervene and better influence the future, but hierarchies in the transport chain are changing. Ports therefore need to be active in extending or even maintaining their hinterlands (Van Klink and van den Berg, 1998; McCalla, 1999; Notteboom and Rodrigue, 2005). Moglia and Sanguineri (2003) analysed the role of a public port authority in the activities of private companies such as terminal operators, particularly in terms of stimulating private investment, for example acquiring land within the port for logistics operations. The authors also highlighted the importance of port authorities having a member on the board of private organisations carrying out commercial activities within the ports. The 'hinterland access regime' proposed by De Langen and Chouly (2004) views the collaborative activities undertaken by a number of actors as a governance issue. The governance issue comes to the fore because port authorities have limited influence on infrastructure development beyond the port perimeter.

Port Regionalisation

Notteboom and Rodrigue (2005) argued that the work of Bird (1963), Taaffe et al. (1963), Hayuth (1981) and Barke (1986) did not address the rising importance of inland load centres to port development, particularly the integration of inland terminals within the transport network. In some ways, the regionalisation concept can be seen as a combination of load centres (Hayuth, 1981) and priority corridors (Taaffe et al., 1963). While the port regionalisation concept has clear antecedents in both Bird (1963) and Taaffe et al. (1963), some empirical applications of port regionalisation in new contexts (Ducruet et al., 2009; Wang and Ducruet, 2012) imply in their focus on inland control that the port regionalisation concept's major debt is to Taaffe et al. (1963). Rimmer and Comtois (2009) are critical not just of the port regionalisation concept (they ask 'what is regionalisation but decentralisation?', p.38) but of what in their view is an overly land-focused approach taken by Taaffe et al. (1963) and subsequent authors. They believe that the increased focus on landside activities in analyses of port development is unwarranted.

The port regionalisation concept is difficult to define concisely, as it describes a somewhat loose collection of activities:

> Port regionalisation thus represents the next stage in port development (imposed on ports by market dynamics), where efficiency is derived with higher levels of integration with inland freight distribution systems ... . Many ports are reaching a stage of regionalisation in which market forces and political influences gradually shape regional load centre networks with varying degrees of formal linkages between the nodes of the observed networks (p.302).

Port regionalisation is thus a term that encapsulates a variety of integration and cooperation strategies, with varying motivations of hinterland capture, control and competition.

In political discourse, the term 'regionalisation' refers to a shift in focus, power or responsibility to the regional level, a process of devolution from the national level. A vast literature exists on issues of 'hollowing out' and 'filling in' of governance capacity (e.g. Goodwin et al., 2005), including critiques of the 'new regionalism' (Lovering, 1999) as much European Union funding is channelled directly from the EU to regions, thus bypassing the national scale. In the port regionalisation discussion, the term is used in the sense that the port's focus moves spatially from a local to a regional focus, a change in emphasis that is reflected in the seeking of new ways of integrating with inland transport systems.

According to Ducruet (2009), the 'port region' has never been defined adequately. Modern ports act as gateways to the trade of larger regions, but they remain embedded in the territorial and economic characteristics of their immediate geographical region. The port region is, therefore, multifaceted, incorporating the local economy, the wider hinterland and the port range. The 'region' in the port regionalisation concept can therefore be understood more accurately as the hinterland, which is itself a fluid concept depending on how integrated a port is with inland transport and logistics networks and whether hinterlands are considered as congruent or extended.

The port regionalisation concept can also be linked to the 'multi-port gateway region' proposed by Notteboom (2010), which defines a number of ports competing to serve an overlapping hinterland. Intermodal connections as well as suitable logistics structures are paramount to capture and control these areas, and the regional differences and specificities of each port region, both within and between continents, will determine how regionalisation plays out in terms of terminals, corridors and institutional relationships. Rodrigue and Notteboom (2010) identified globalisation, economic integration and intermodal transport as three major influences on what they call 'the regionalism of freight distribution' (p.498), concluding that 'regionalism results in different strategies' (p.504), and they have noted elsewhere that 'there is no single strategy in terms of modal preferences as the regional effect remains fundamental' (Notteboom and Rodrigue, 2009a: p.2).

Once the importance of this 'regional effect' is recognised, attempting to capture processes of 'regionalism' within a single concept of 'regionalisation' must of necessity be difficult, in particular when recognising the unreliable relationship between region and hinterland. The research in this book will derive the key elements of the port regionalisation concept in order to examine in more detail how they reveal such regional effects or 'regionalism' through the spatial and institutional characteristics of each activity.

The preceding discussion highlights that the port regionalisation concept has both spatial and institutional aspects. Spatial refers to physical developments such as terminals and rail/barge corridors. An essential component of terminals and corridors is market capture which is not based on physical developments but institutional relationships. In most cases the infrastructure is common-user so the port authority or terminal cannot control the physical corridors but rather focuses on operational agreements with transport and logistics providers. The goal of the port actor is for the traffic to come through its port, regardless of who the transport operator is, what mode they use and which corridor they follow.

As shown above, the early port development literature was focused more on spatial development than actor-centric approaches, due in part to the historical industry structure. While recent literature makes more of the distinction between port actors, insufficient attention has been given to the identification of different strategies. Notteboom and Rodrigue (2005) assert that regionalisation is imposed on ports but the dynamics of this concept are unclear, such as the determinants of the 'varying degrees of formal linkages' (p.302) and the way that 'market forces and political influences' (p.302) affect the processes of integration and collaboration required for successful port regionalisation. While one study cannot cover all possibilities, three key aspects will be selected for further study.

Deriving the Three Key Aspects of Intermodal Transport and Logistics From the Port Regionalisation Concept

The first key distinction of the port regionalisation concept is its focus on inland terminals. Notteboom and Rodrigue (2005) state that the concept 'incorporate[s] inland freight distribution centres and terminals as active nodes in shaping load centre development' (p.299), and that this process is 'characterised by strong functional interdependency and even joint development of a specific load centre and (selected) multimodal logistics platforms in its hinterland' (p.300). The authors also propose that the regionalisation phase 'promote[s] the formation of discontinuous hinterlands' (p.302) suggesting that 'a port might intrude in the natural hinterland of competing ports' (p.303) by using an inland terminal as an 'island formation' (p.303).

The second key aspect of the concept is the role of the market, in particular the changing nature of logistics operations. Notteboom and Rodrigue (2005) state that 'regionalisation results from logistics decisions and subsequent actions of shippers and third-party logistics providers' (p.306), and that 'the transition towards the port regionalisation phase is a gradual and market-driven process, imposed on ports, that mirrors the increased focus of market players on logistics integration' (p.301). They go on to note that 'logistics integration ... requires responses and the formulation of strategies concerning inland freight circulation. The responses to these challenges go beyond the traditional perspectives centered on the port itself' (p.302).

Thirdly, Notteboom and Rodrigue (2005) discuss proactive attempts to influence load centre or inland terminal development, as something of an alternative to their suggestion that port regionalisation is 'imposed on ports' (p.301). They state that 'the trend towards spatial (de)concentration of logistics sites in many cases occurs spontaneously as the result of a slow, market-driven process. But also national, regional and/or local authorities try to direct this process by means of offering financial incentives or by reserving land for future logistics development' (p.306). Yet they warn against the danger of optimism bias: 'a lack of clear insights into market dynamics could lead to wishful thinking by local governments ... . This can lead to overcapacity situations' (p.307). While the authors suggest that ports should not 'act as passive players' (p.306), and should adopt 'appropriate port governance structures' (p.306) to deal with these new challenges, they state clearly that 'the port itself is not the chief motivator for and instigator of regionalisation' (p.306). However, they note that 'the port authority can be a catalyst even when its direct impact on cargo flows is limited' (p.307). They observe that many different types of relationships can be developed between the port and the inland actors, depending largely on 'the institutional and legal status of the partners involved' (p.307). Essential to an understanding of this aspect is the uneven distribution of costs and benefits resulting in a free rider problem: 'Port authorities are generally aware that free-rider problems do exist. This might make port authorities less eager to embark on direct formal strategic partnerships with a selected number of inland terminals. Instead, port authorities typically favour forms of indirect cooperation ... which are less binding and require less financial means' (p.310) as 'a seaport cannot make cargo generated by an inland terminal captive to the port' (p.310).

If port regionalisation is proposed as an observable stage in port development theory, the implication is that its presence is possible or even probable in the majority of instances. This research will look at the three key aspects just addressed in order to identify how they enable or constrain these integration processes embedded within the port regionalisation concept. This is not to state that port regionalisation is not happening, but rather to pay more attention to how it occurs and in particular what barriers exist. The research questions for this study relate to each of these three aspects:

1. How can different strategies of inland terminal development influence port regionalisation processes?
2. How can logistics integration and inland freight circulation influence port regionalisation processes?
3. How can collective action problems influence port regionalisation processes?

These questions are deliberately open-ended as the aim of the research is to learn about the processes of which the port regionalisation concept is comprised. In each empirical chapter the appropriate literature will be examined to determine the main features of each of the three questions. Case studies will be selected through which to describe and discuss how the integration processes as defined within the port regionalisation concept are enabled or constrained, revealing the spatial and institutional characteristics of intermodal transport and logistics.

Conclusion

The literature review in each empirical chapter will address each of these three topics in order to develop the key factors for the study, appropriate to focus each case. These factors (with sub-factors where relevant) will be used to guide the data collection and analysis, within an inductive methodology as described in Chapter 1. For each of the three empirical chapters, a matrix based on these factors will form the basis of the research process, producing findings that can be used to answer each of the three research questions.

Once these questions have been addressed, the findings can then contribute to a discussion on the link between the spatial and institutional characteristics of intermodal transport and logistics, in which spatial models of port and hinterland development are expanded via an institutional analysis of the role of governance. In order to facilitate this analysis, the following chapter will review the institutional literature, with a particular focus on governance in transport and logistics.

Chapter 3
The Role of Institutions in Intermodal Transport and Logistics

Introduction

This chapter covers the role of institutions, beginning with general applications in sociology, moving through the 'institutional turn' in geography and then introducing the specific context of institutions and organisations in transport and logistics, which focuses to a large degree on issues of governance. The argument is made that unsuccessful intermodal transport developments result from the fact that, while the operational realities of intermodal transport are relatively well known, the institutional challenges are less well understood. The key issues are outlined in order to demonstrate why such an approach is necessary to explore the challenges to intermodal transport and logistics established in Chapter 2. The key governance issues are then derived, which will be used to situate the empirical findings in the discussion in Chapter 7.

An Introduction to Institutionalism[1]

Institutionalism developed out of neoclassical economics due to an increasing focus on the social, cultural and historical context of economic events rather than what was viewed as an overly theoretical and non-contextual framework of universal laws. Jaccoby (1990) identified four movements: from determinacy to indeterminacy, from endogenous to exogenous determination of preferences, from simplifying assumptions to behavioural realism, and from synchronic to diachronic analysis.

According to Coase (1983), this early form of institutional economics was not theoretical enough to prevail against neoclassical approaches. Scott (2008) suggested that 'new institutionalism' in the social sciences is actually the direct descendent of this early or 'old' institutional economics, whereas what became known as new institutional economics (NIE) is closer to the original (and still prevailing) neoclassical economics. NIE tends to operate within the neoclassical view, in which the firm behaves rationally by acting in certain ways to reduce transaction costs (Jessop, 2001), although it has departed from some neoclassical assumptions, such as perfect information and costless transactions (Rafiqui, 2009).

---

1   This section draws on the literature review in Monios and Lambert (2013a).

The term 'new institutional economics' was first used by Williamson (1975). It was developed out of Coase's (1937) work on transaction costs, which are those costs incurred when dealing with a separate firm through the price mechanism. For instance, if two firms merge then the previously external costs of doing business will be internalised. Neo-institutional economists utilise the theory of the firm in order to examine different methods of lowering transaction costs such as mergers, alliances and contracts.

In maritime transport studies NIE has been used by some authors to explore different methods of coordinating hinterland transport chains (e.g. de Langen and Chouly, 2004; Van der Horst and de Langen, 2008; Van der Horst and Van der Lugt, 2009). Institutional geography, on the other hand, examines how these structures vary across space, place and scale (e.g. Hall, 2003; Jacobs, 2007; Ng and Pallis, 2010).

One of the key writers on institutionalism is North (1990), for whom institutions represent the rules of the game, whereas organisations are the players. These issues arise most pertinently when attempting to define the state through its organisations and institutions. Jessop (1990: p.267) defined the state as a 'specific institutional ensemble with multiple boundaries, no institutional fixity and no pre-given formal or substantive unity'. Government influence or capacity to innovate is embedded in both formal and informal institutions, which, according to González and Healey (2005: p.2059), are 'located in the practices through which governance relations are played out and not only in the formal rules and allocation of competences for collective action as defined by government laws and procedures'.

Taking a comparable perspective, Aoki (2007) identified exogenous and endogenous institutions. The former represent the rules of the game (following North, 1990), while the latter characterise the equilibrium outcome of the game. Aoki (2007: p.6) combined both elements in the following definition: 'An institution is a self-sustaining, salient pattern of social interactions, as represented by meaningful rules that every agent knows and are incorporated as agents' shared beliefs about how the game is played and to be played'.

Legitimacy is a key concept for a successful organisation, and it is derived from its relation to institutions. Suchman (1995: p.574) defined legitimacy as 'a generalised perception or assumption that the actions of an entity are desirable, proper or appropriate within some socially constructed system of norms, values, beliefs and definitions'. Yet Meyer and Rowan (1977) traced a conflict between legitimacy and efficiency. They argued that organisations adopt formal structures in order to achieve legitimacy rather than out of any practical requirement arising naturally from their operations. Indeed, such formal structures may even decrease efficiency. They went further to insist on a divergence between the formal structure of an organisation and its day-to-day activities. The result of this divergence is that innovation may be stifled by inappropriate formal structures, and monitoring may become primarily ceremonial and related to the formal structure rather than to the real activities of the organisation.

Transferring a governance structure from elsewhere can be problematic (Ng and Pallis, 2010). Meyer and Rowan's (1977) description of the creation of new organisations has a great deal of relevance for modern organisational design, particularly when transferring a governance structure from one scale or space to another: 'The building blocks for organizations come to be littered around the societal landscape; it takes only a little entrepreneurial energy to assemble them into a structure. And because these building blocks are considered proper, adequate, rational and necessary, organizations must incorporate them to avoid illegitimacy' (p.345). Furthermore, the process engenders more of the same kind of structural legitimacy, which may be observed in the rise in the number of quangos (quasi-non-governmental organisations). The authors noted that 'institutionalized rationality becomes a myth with explosive organizing potential' (p.346).

The constant changing and re-making of institutions is an ongoing source of difficulty. Jessop (2001) identified 'the contingently necessary incompleteness, provisional nature and instability of attempts to govern or guide them' (p.1230). This problem has been defined elsewhere as multi-scaled governance (Hooghe and Marks, 2001), which results in complications through confused sovereignty, multiple authorities and funding sources (Meyer and Scott, 1983; Scott and Meyer, 1983).

Moe (1990: p.228) observed that political organisations are compelled to make trade-offs that economic organisations do not:

> [Political organizations] are threatened by political uncertainty. They want their organizations to be effective, and they also want to control them; but they do not have the luxury of designing them for effectiveness and control. Economic decision-makers do have this luxury – because their property rights are guaranteed. They get to keep what they create.

As a result, the structures of political organisations arise from the interaction between voters (or other political interest groups), politicians and the civil service. An attractive strategy is, therefore, 'not to try to control how it gets exercised over time, but instead to limit it ex ante through detailed formal requirements. ... In politics, it is rational for social actors to fear one another, to fear the state, and to use structure to protect themselves – even though it may hobble the agencies that are supposed to be serving them' (Moe, 1990: p.235). The way governments channel money towards infrastructure investment to the benefit of private firms can be viewed in this light. As will be noted in the case studies in this book, public agencies establish complex funding and grant structures so that any decisions are based on rules established at the start rather than being the decision of individual politicians or administrators. The result can be, as Moe (1990) says, a 'hobbled' ability to wield effective influence.

Path dependency is a key issue in the economic geography literature; it arises from high setup costs, learning effects, coordination effects and adaptive expectations and can lead to indeterminacy, inefficiencies, lock-in and the primacy of early events (David, 1985; Arthur, 1994). Martin (2000) noted how institutions 'tend to evolve incrementally in a self-reproducing and continuity-preserving way' (p.80) and also highlighted the importance of different development paths of institutions at different regional and local contexts: 'if institutional path dependence matters, it matters in different ways in different places: institutional-economic path dependence is itself place-dependent' (p.80).

States at all levels experience increasing pressure to establish an entrepreneurial culture able to draw progressively mobile global capital flows to their region, but the identification of scales is important because 'the capital-labour nexus was nationally regulated but the circulation of capital spiralled out to encompass ever-larger spatial scales' (Swyngedouw, 2000: p.69). Due to the decreasing role of the national state, local and regional authorities try to secure these flows through strategies of clustering and agglomeration that have been observed to be successful in some cases, although it is unclear whether the clustering is the cause or effect of a strong institutional setting.

The concept of 'institutional thickness' was proposed by Amin and Thrift (1994, 1995), defined as a measure of the quality of an institutional setting. The authors identified four elements: a strong institutional presence; a high level of interaction among these institutions; a well-defined structure of domination, coalition building and networking; and the emergence of a common sense of purpose and shared agenda. The concept has not been applied widely, but where it has been attempted the focus has been almost exclusively on economic development (see Raco, 1998, 1999; Henry and Pinch, 2001). Henry and Pinch (2001) identified a coalescence between the rise of institutionalism as a subject within economic geography and the growth of the 'new regionalism' as a focus on regional economic development.

MacLeod (1997, 2001) noted how 'institutional thickness' shares similar ground with other concepts such as Lipietz's (1994) 'regional armature', Cooke and Morgan's (1998) 'institutions of innovation' and Storper's (1997) 'institutions of the learning economy'. He demonstrated the institutional density of lowland Scotland which therefore represented a good case of institutional thickness. Indeed, he noted that lowland Scotland has potentially achieved 'institutional overkill' by establishing too many organisations: 'These processes help to illustrate that, as Amin outlines, attempts to achieve collaboration between entrepreneurs and institutions through policy dictate and 'overnight institution building' can be deeply problematic (Amin, 1994)' (p.308). MacLeod noted that this institutional thickness had not helped Scotland retain transnational capital, nor develop new Scottish-controlled industry, leading him to conclude that one must be careful when de-emphasising the role of the nation-state.

MacLeod (2001) further demonstrated the necessity of taking a multiscalar perspective on the state, 'so as to reveal which particular regulatory practices and elements of an 'institutional thickness' are scaled at which particular level .... These spatial and scalar selectivities (Jones, 1997) can occur through state-run policies like defence or through targeted urban and regional policies' (p.1159). This ongoing process cannot be accepted uncritically as an input into an institutional analysis: 'far from being existentially given, geographical demarcations such as cities and regions are politically constructed stakes in a perpetual sociospatial struggle over capitalist relations and regulatory capacities' (p.1159). Similarly, Amin (2001: p.375) added that it is 'the management of the region's wider connectivity that is of prime importance, rather than its intrinsic supply-side qualities'.

The institutional thickness concept was applied by Pemberton (2000) to a study of transport governance in the northeast of England, who followed Jessop's use of neo-Gramscian state theory, as a way to include the role of the state as advocated by MacLeod (2001). Coulson and Ferrario (2007) questioned the lack of penetration of institutional thickness as a critical approach over the last decade, resulting in an identification of potential issues with cause and effect, a risk of conflating organisations with institutions and the difficulty of creating or replicating an institutional structure through policy actions.

The key elements of institutional analysis can be summarised as a potential conflict between an organisation's legitimacy and its efficiency or agency (Meyer and Rowan, 1977; Monios and Lambert, 2013a), difficulties in transferring a governance structure from one institutional setting to another (Meyer and Rowan, 1977; Ng and Pallis, 2010), the constant changing and re-making of institutions (Jessop, 2001), scale issues leading to complications through confused sovereignty, multiple authorities and funding sources (Meyer and Scott, 1983; Scott and Meyer, 1983) and the path-dependent trajectory of institutional development (David, 1985; Arthur, 1994; Martin, 2000). One must also recognise not only formal but informal institutions (González and Healey, 2005). Finally, Rodríguez-Pose (2013) warns of the difficulty of measuring institutional influence (especially informal institutions) and the related difficulty of instituting such desirable influence through policy.

Institutions and Governance

Much institutional literature has focused on the issue of governance, which can be defined very simply as an act or process of governing. It has in the past been used interchangeably with government, but in the last two decades governance rather than government has become the preferred term. As power is devolved from governments to other bodies and representation of other interests is increased, official government institutions become only one part of the totality of the governance process (Romein et al., 2003; Jordan et al., 2005). Governance must then be understood as a process of distributing authority and allocating resources, of managing relationships, behaviour or processes to achieve a desired outcome.

Taking this perspective of governance as a process, the state becomes 'merely an institutional ensemble; it has only a set of institutional capacities and liabilities which mediate that power; the power of the state is the power of the social forces acting in and through the state' (Jessop, 1990: pp.269–70). Brenner (1999: p.53) describes the state as a 'polymorphic multiscalar institutional mosaic', within which, according to Swyngedouw (1997: p.141), spatial scales are 'perpetually redefined, contested and restructured in terms of their extent, content, relative importance and interrelations'. They are 'a series of open, discontinuous spaces constituted by the social relationships which stretch across them in a variety of ways' (Allen et al., 1998: p.5).

As territorial political boundaries become less important, the relational element of governance is foregrounded. Political structures may remain ostensibly linked to territorial spaces (e.g. physical boundaries), but their legitimacy and agency are relationally constructed, through the power of regional elites and industry players (MacLeod, 1997; Allen and Cochrane, 2007; Monios and Wilmsmeier, 2012b). Governance, therefore, becomes increasingly about working across boundaries, between government organisations, non-government organisations and individuals, as well as incorporating multiple scales of government (Marks, 1993; Hooghe and Marks, 2003). This process can be linked to recent trends towards decentralisation and devolution (Peck, 2001; Rodríguez-Pose and Gill, 2003), which nonetheless are not necessarily an actual transfer of power but more of a qualitative restructuring (Brenner, 2004), characterised as uneven processes of hollowing out (Rhodes, 1994) and filling in (Jones et al., 2005; Goodwin et al., 2005) that can result in asymmetrical acting capacity.

The changing role of political institutions is a key topic, but, more than simply their formal boundaries and powers, much governance literature addresses the process, asking questions about how power should be exercised, performance measured and outcomes regulated. How such processes are enacted is the core of the difference between governance and government. What is at stake is not the location of official responsibility but how a process is governed and an outcome achieved. These outcomes cover policy areas such as climate change, resource management, transport provision, accessibility and social inclusion. Effective governance can limit damage and protect social rights by regulating access to an environment, whether that be regulating access of mining companies to protect water quality or regulating car use to reduce local air pollution. In addition to considering the governance model most likely to achieve a specific political outcome, the outcome itself must also be considered. This means that effective governance is not always measured by, for example, a measured reduction in an undesirable outcome such as pollution. Governance reform can be pursued to increase the representation of minority stakeholders, or to improve transparency and accountability in decision making. Where governance and institutional approaches have been applied to (passenger) transport, the interest has been predominantly to transport provision and its regulation by government organisations (Stough and Rietveld, 1997; Pemberton, 2000; Gifford and Stalebrink, 2002; Geerlings and Stead, 2003; Marsden and Rye, 2010; Curtis and Lowe, 2012; Legacy et al., 2012). Governance theory has been applied in the field of freight transport to assess the role of multi-level governance in the regulation of shipping policy; this process involves actors at international, supranational, national, regional and local levels (Pallis, 2006; Roe, 2007, 2009; Verhoeven, 2009). The major application of governance theory to freight transport, however, has been to port governance.

Governance Applied to Ports

Following on from the discussion in Chapter 2, it is clear that control of ports is a significant lever for governments to manage trade and its attendant economic benefits. The literature shows that, over recent decades, a general trend has been observed for port management to move from the public to the private sector. Different models of port governance have been the subject of considerable research (e.g. Everett and Robinson, 1998; Baird, 2000, 2002; Hoffman, 2001; Baltazar and Brooks, 2001; Cullinane and Song, 2002; Brooks, 2004; Brooks and Cullinane, 2007; Pallis and Syriopoulos, 2007; Brooks and Pallis, 2008; Ferrari and Musso, 2011; Verhoeven and Vanoutrive, 2012). Four models of port governance were classified by the World Bank (2001, 2007): the public service port, the private port, the tool port (a mixed model where private sector operators perform some of the operations but under the direction of public sector managers) and the landlord port (the public sector retains ownership while the terminal management and operations are leased to private sector operators). The landlord model has become increasingly common across the globe, and has indeed been encouraged by the World Bank and others, but implementation of port devolution policies has been observed to vary according to local conditions (e.g. Baird, 2002; Wang and Slack, 2004; Wang et al., 2004; Ng and Pallis, 2010).

The ongoing reform of port governance requires a focus on the specifics of various processes in which a port actor might engage. Several topics have been addressed, such as the influence of shipping networks (Wilmsmeier and Notteboom, 2011), the role of the port authority in the cluster of associated businesses and services agglomerated around a port (Hall, 2003; de Langen, 2004; Bichou and Gray, 2005; Hall and Jacobs, 2010), the development of new competencies such as hinterland investment (Notteboom et al., 2013), port competition (Jacobs, 2007; Ng and Pallis, 2010; Sanchez and Wilmsmeier, 2010; Jacobs and Notteboom, 2011; Wang et al., 2012) and the devolution of port governance from one level of government to another rather than from the public to the private sector (Debrie et al., 2007).

The advantages of greater private sector involvement in ports are related primarily to increased efficiency and reduced cost to the public sector. Negative impacts include the loss or increased ambiguity of state control as well as the difficulties and risks involved in managing the tender process and subsequent monitoring (Baird, 2002). It has also been proposed that governance decisions are not always related to port performance (Brooks and Pallis, 2008). Debrie et al. (2013) argued for a deeper contextualisation of port governance models, adding a spatial element by combining the institutional context (relationship between public and private actors and relative decision-making powers) with characteristics of the local market and societal and cultural factors impacting on motivations for public intervention. This kind of contextualisation is essential to governance analyses because applying a generic governance model in different local settings can lead to asymmetric results (Ng and Pallis, 2010). The diversity of port functions (see also Beresford et al., 2004; Sanchez and Wilmsmeier, 2010) is why, according to Bichou and Gray (2005), simple taxonomies are difficult to create; the suggestion is, therefore, that three elements should be included: the role of public and private actors, the governance model and the scope of facilities, assets and services. This approach will be used in Chapter 7 to expand simple terminal governance models derived from the literature with a strong operational component sourced from the empirical chapters.

Governance Applied to Intermodal Transport

Traffic must be consolidated in order to support the economic viability of intermodal corridors. A result of this need is the increasing tendency for the logistics operations at an inland freight facility to be presumed to operate in conjunction with the transport activities. Earlier, intermodal terminals had been the main focus of the literature, with logistics platforms being addressed separately in the logistics literature. Confusion has arisen in the transport literature in recent years as both transport and supply chain functions have been discussed interchangeably, without addressing the governance issue and how the two functions and physical spaces relate to each other. This problem forms the motivation for the classification developed in Chapter 7 from an analysis of the empirical cases in this book.

The topic of intermodal transport in general and intermodal terminals in particular has been increasingly popular in the literature over the last five years, but governance has rarely been addressed directly. This may be because inland freight nodes tend to be smaller concerns than ports, with simpler governance structures and less government involvement. Some landlord models are in evidence, although, unlike with ports, government involvement in inland freight facilities is more likely in the start-up phase using public money to attract a private operator into the market. The hope is commonly that, following successful development, the site will be run by private operators with no further government involvement (although the case analysis in this book demonstrates that there are exceptions).

One of the few direct applications of governance to inland terminals was by Beresford et al. (2012), who applied the World Bank port governance model (public, tool, landlord, private) to dry ports. This was a useful approach facilitating an analysis of the relationship between the owner and operator. Beresford et al. (2012) also drew on the UN-ESCAP (2006) concentric model, in which the middle ring contains the container yard and container freight station, expanding out to a container depot, then the third ring is for logistics and finally an outer ring for related processing and industrial activities on the periphery of the area. Similarly, in the three-stage concentric model of Rodrigue et al. (2010), the intermodal terminal is at the centre of the activity, a larger ring includes any logistics activities that may or may not be part of the same site, and finally a third level accounts for any wider retail and manufacturing activities in the hinterland that may be loosely related to the site.

Concentric representations can be misleading; they tend to imply that the intermodal terminal is situated at the heart of a unified logistics platform. In reality, the terminal will be found at the edge of the site (see Chapter 2) and will primarily serve customers external to the logistics platform. The concentric model also masks the reality that, in most cases, the terminal(s) are separate to the logistics platform. Even if the terminal and the logistics platform are located in close proximity to each other, they will still require entry and exit via a separate gate entailing appropriate security operations. Indeed, they are more likely to be located a few miles away from each other (thus requiring an additional road haul), or otherwise the terminal may be located in an area with several logistics operations of varying sizes, types and specialisations, which may or may not have transport requirements suitable for intermodal transport.

Governance Applied to Logistics

Analysis of governance in the logistics literature relates to logistics integration, which is a subset of the wider topic of supply chain integration. Numerous motivations have been identified in the literature for supply chain integration; these include cost reduction through efficiency advances, resource complementarity, customer requirements, technology adoption, changes in supply chain partners and structure and competitive pressures. Potential challenges have also been acknowledged, such as lack of top management support, misaligned incentives, lack of trust, lack of information sharing, inconsistent performance measures and lack of joint decision making (Whipple and Frankel, 2000; Min et al., 2005; Simatupang and Sridharan, 2005; Cruijssen et al., 2007; Fawcett et al., 2008a, 2008b; Richey et al., 2010; Guan and Rehme, 2012). An important lesson from this literature that should be applied to transport analysis is that internal logistics integration is required as well as external. For example, integration of planning between the logistics and purchasing department is necessary if the logistics department is attempting to integrate services with external organisations (Stank et al., 2001; Gimenez and Ventura, 2005; Lambert et al., 2008; Chen et al., 2009).

Governance models have been analysed in far greater detail in the supply chain literature than in transport, with a particular background in theoretical approaches such as transaction cost economics (see Chapter 6 for applications of these theories to transport). Strategies adopted in supply chain management stretch from a purely transaction- or market-based approach at one end to a fully integrated or hierarchical ownership model at the other (Golicic and Mentzer, 2006; Rinehart et al., 2004). Market-based models are governed by contracts of varying duration, regularly compared with the price and service offered by competitors, whereas integration strategies may result in an outright purchase or merger of one firm by another or the creation of a new organisation through a joint venture. A variety of dynamic hybrid or relational models fit in between these two extremes, such as written contracts without equity involvement and minority stake agreements (Williamson, 1975; Parkhe, 1991; Dussauge and Garrette, 1997; Klint and Sjöberg, 2003; Rinehart et al., 2004; Todeva and Knoke, 2005; Halldorsson and Skjøtt-Larsen, 2006; Humphries et al., 2007; Schmoltzi and Wallenburg, 2011). These models can be classified according to the equity stake, but of more relevance to this study is that they can also be characterised by increasingly integrated services, from basic cooperation to coordinating business planning to strategic long-term process collaboration (Spekman et al., 1998; Lambert et al., 1999; Whipple and Russell, 2007). When analysing the relation between intermodal terminals, logistics platforms and intermodal corridors, ownership or equity investment is only part of the story; the different kinds and levels of process integration often determine whether sufficient traffic will be consolidated on the intermodal corridor which is necessary to unpin the viability of any site development.

Market or contractual governance requires relationships to be managed through contracts, which will entail incentives and penalties. As firms move towards greater collaboration, relational characteristics such as trust, information sharing and mutual decision-making become more important. Partner relationships can be coordinated through several mechanisms, such as monitoring, incentives/hostages and social enforcement based on personal relationships (Dyer and Singh, 1998; Wathne and Heide, 2000; Hernández-Espallardo and Arcas-Lario, 2003).

The preceding discussion showed that, in transport, governance is about coordination of service requirements. By contrast, in logistics and supply chain management the focus is on firm creation (Wilding and Humphries, 2006)[2] and resource utilisation (Schmoltzi and Wallenburg, 2011),[3] leading eventually to a relational or network approach (Dyer and Singh, 1998; Pfohl and Buse, 2000; Skjoett-Larsen, 2000; Zacharia et al., 2011).

Bowersox et al. (1989) established a five-stage model of logistics integration:

1. Single transactions;
2. Repeated transactions;
3. Partnerships;
4. Third-party agreements;
5. Integrated logistics service agreements.

According to this model, the partnership stage is when the shipper retains control of planning and management, while a third-party agreement is when the 3PL takes a more direct role in the relationship with a tailored service requiring information sharing, which increases the level of trust required. Finally, an integrated service agreement is where the entire logistics function or at least

---

2   For transaction cost economics see Coase (1937), Williamson (1975, 1985).

3   For more on the resource-based view see Wernerfelt (1984), Barney (1991), Dyer and Singh (1998), Lavie (2006), Hernández-Espallardo et al. (2010), Peters et al. (2011).

large parts of it have been outsourced to the 3PL. This will necessarily require a higher level of information integration possibly through joint ICT, and may also include additional value-added services as the inventory may in fact be stored at warehouses operated by the 3PL. An 'organisational implant' (Grawe et al., 2012) may be used, which is when a representative from a 3PL is placed within the client organisation.

Summarising the Key Governance Issues Relevant for Intermodal Transport and Logistics

The governance models examined in the preceding discussion of governance in intermodal transport and logistics highlight the relation between the owner and operator and the separation of the transport function from the logistics function, as well as the role played by the trade-related activities at the periphery. They thus provide a useful beginning; they do not, however, disaggregate and identify the different kinds of relations between each level. The governance literature highlights the importance of working across boundaries and achieving cooperation among various interests and voices, and the supply chain literature explicitly requires consideration of internal and external integration processes. These elements must be incorporated in an analysis of the governance of intermodal transport and logistics.

The primary issues can be grouped within the following four categories:

1. Roles of the public and private sectors in processes of planning and development.
2. The relation between the original developer and the eventual operator, including selling and leasing.
3. The relationship between the transport and logistics functions, and other issues internal to the site.
4. The site functions and operational model, including the relationships with clients and external stakeholders.

Pure cost analysis as the basis for development of intermodal infrastructure and operational subsidies can be misleading because many organisational and institutional difficulties can prevent the efficiency of operations necessary to compete economically with road haulage. In order to achieve the ideal outcomes of full trains and high usage, all the relevant organisations must play their part in the smooth planning, development, scheduling and eventual operation of the intermodal chain, which includes the port, corridor, terminal, rail operations and the entire logistics system in which they are situated. The aim of this research is to identify the key institutional relationships requiring greater investigation.

Conclusion

The following three chapters will present empirical case studies of intermodal terminals, logistics and corridors, respectively. Each empirical chapter will begin with a review of the relevant literature to identify the key topics for analysis. The findings of each of these three chapters will then be brought together in an institutional analysis in Chapter 7 based on the four factors identified in this chapter.

Chapter 4
Case Study (Europe): Intermodal Terminals

Introduction

Chapter 4 is the first of three empirical chapters. A sample of 11 European intermodal terminals is described and compared, divided into one group with port investment and one group without, in order to identify different models of inland terminal development. Their key features and functions are discussed, including the roles of the public and private sectors, relations with ports and rail operators and their situation within the logistics sector. The case comparison reveals that inland terminals developed by landside actors often experience a conflict of strategy with port actors (either port authorities or terminal operators). Port actors have difficulty acting beyond the port perimeter but some port terminal operators have begun to demonstrate successful investments in inland terminals due to their increased involvement in the development of intermodal services for managing their container throughput. The difficulties of successful terminal development unless embedded firmly within a rail, port or logistics model are demonstrated.

A Review of the Literature on Intermodal Terminals[1]

The Spatial Development of Inland Terminals

As noted in Chapter 2, Notteboom and Rodrigue (2005) stated that the port regionalisation concept 'incorporate[s] inland freight distribution centres and terminals as active nodes in shaping load centre development' (p.299), and that this process is 'characterised by strong functional interdependency and even joint development of a specific load centre and (selected) multimodal logistics platforms in its hinterland' (p.300). The authors also propose that the regionalisation phase 'promote[s] the formation of discontinuous hinterlands' (p.302) suggesting that 'a port might intrude in the natural hinterland of competing ports' (p.303) by using an inland terminal as an 'island formation' (p.303). While location splitting via spatially discontinuous inland development of ports has begun to be treated in the theoretical literature (Cullinane and Wilmsmeier, 2011), the role of the inland terminal in this spatial theory has been lacking. This chapter will extend the understanding of port regionalisation by exploring its relation to different types of inland terminal

---

1 This literature analysis draws on work by the author published in Wilmsmeier et al. (2011) and Monios and Wilmsmeier (2012a).

concepts; in this way the under-theorised area of inland terminals can be tied into the established theoretical grounding of the spatial development of ports.

Classification of inland freight facilities and the activities in which they engage is difficult, and, despite some earlier analysis of their functions and locations (Hayuth, 1980; Slack, 1990; Wiegmans et al., 1999), it is only in recent years that their spatial and institutional characteristics have begun to be treated in detail. Rodrigue et al. (2010: p.2) asserted that 'while a port is an obligatory node for the maritime/land interface, albeit with some level of inter-port competition, the inland port is only an option for inland freight distribution that is more suitable as long as a set of favourable commercial conditions are maintained'. Similarly, Notteboom and Rodrigue (2009: p.2) stated that 'there is no single strategy in terms of modal preferences as the regional effect remains fundamental. Each inland port remains the outcome of the considerations of a transport geography pertaining to modal availability and efficiency, market function and intensity as well as the regulatory framework and governance'.

Notteboom and Rodrigue (2009) suggested that it is impossible to have firm definitions as each site is different, therefore it is best to focus on the key aspects of each. Rodrigue et al. (2010) related the multiplicity of terms to the variety of geographical settings, functions, regulatory settings and the related range of relevant actors, and proposed that the key distinction is between transport functions (e.g. transloading between modes, satellite overspill terminals or load centres) and supply chain functions (e.g. storage, processing, value-added). This functional approach is similar to the Roso et al. (2009) distinction between close, mid-range and distant terminals, and the later seaport-based, city-based and border-based model proposed by Beresford et al. (2012), as both of these tripartite divisions are distinguished by the typical functions of each node.

Table 4.1 lists the inland terminal classifications found in the literature.

Table 4.1   Inland freight node taxonomies

No.	Name	Description
1	Inland clearance (or container) depot	The focus here is on the ability to clear customs at the inland origin/ destination site rather than at the port. Started to spring up in the 1960s. Therefore some kind of warehouse area (could just be small) with customs clearance. Any transport mode is acceptable within this definition. See Hayuth (1980); Beresford and Dubey (1991); Garnwa et al. (2009); Jaržemskis and Vasiliauskas (2007); Pettit and Beresford (2009).
2	Container freight station	This is basically a shed for container stuffing/stripping/ (de-)consolidation. It is not a node in itself but more of a service that may be provided within a port or an inland site.
3	Dry port 1	Synonymous with ICD, either in a landlocked country or one that has its own seaports (see Beresford and Dubey, 1991; Garnwa et al., 2009).

No.	Name	Description
4	Inland port	Favoured in the USA (see Rodrigue et al., 2010). Customs is less of an issue in the USA because 89 per cent of their freight is domestic. As the railroads run on their own private track, terminals are also private nodes, so the management of containers is a closed system for that firm to manage the flow. Some reservations to using it in Europe because there an inland port generally has water access, and in any case inland terminals are not normally the massive gateway nodes that they are in the USA (i.e. fewer than 100,000 lifts annually vs many times that in the USA).
5	Intermodal terminal	Generic term for an intermodal interchange, i.e. road/rail, road/barge. Could strictly speaking be just the terminal with no services or storage nearby, but would generally involve such services. Also referred to as transmodal centre by Rodrigue et al. (2010), which draws attention to its primary function, which is interchange rather than servicing an O/D market but in practice would presumably do some O/D freight as well to make the site more feasible.
6	Freight village, logistics platform, interporto, GVZ, ZAL, distripark (if located in or near a port)	These are big sites with many sheds for warehousing, logistics, etc. and usually relevant services too. May have intermodal terminal or may be road only. May have customs or may not. Distripark is used to denote a site based within or on the outskirts of a port. (Notteboom and Rodrigue, 2009a; Pettit and Beresford, 2009).
7	Extended gate	Specific kind of intermodal service whereby the port and the inland node are operated by the same operator, managing container flows within a closed system, thus achieving greater efficiency and the shipper can leave or pick up the container at the inland node just as with a port. See Van Klink (2000); Rodrigue and Notteboom (2009); Roso et al., (2009); Veenstra et al. (2012).
8	Dry port 2	New definition by Roso et al. (2009). This would seem to be an ICD with large logistics area and intermodal (rail or barge) connection to the port, in combination with extended gate functionality, thus providing an integrated intermodal container handling service between the port and the fully-serviced inland node.
9	Satellite terminal	See Slack (1999). Usually a close-distance overspill site, operated almost as if it is part of the port. Could be considered a short-distance extended gate concept. This should really be rail-connected but some sites are linked by road shuttles (that would seem to ignore the main function which is to overcome congestion, but it can reduce congestion by reducing the time each truck spends in the port on administrative matters).
10	Load centre	This concept could apply to inland terminals or ports, but in the case of the former it refers to a large inland terminal to service a large region of production or consumption. Probably the classic kind of inland node as it serves as a gateway to a large region. Tends to fit well with the American inland port typology. It normally refers specifically to the terminal but generally in this sort of location one would expect to have a lot of warehousing, etc. in the area if not part of the actual site. See Slack (1990); Notteboom and Rodrigue (2005); Rodrigue and Notteboom (2009).

It can be seen from Table 4.1 that, as Rodrigue et al. (2010) argue, inland freight nodes can be divided into two key aspects: the transport function and the supply chain function, with each classification exhibiting various aspects of each. Categories 5, 7 and 8 in Table 4.1 specifically require an intermodal transport connection, while all the others relate to other functions, such as customs, warehousing, consolidation, logistics and other supply chain activities. In practice, many of these sites would have intermodal connections but it is not specifically required within the categorisation.

Relating these issues to the role of the inland terminal in the port regionalisation concept, the key issues are its ability to be an 'active node', to 'impose on ports' and to reflect a focus on 'logistics integration'. The literature review will now look in more detail at how these issues are reflected in recent classifications.

Dry Ports and Extended Gates

Depending on which services they offer, freight handling nodes can be grouped under different categories, as noted in Table 4.1. Some facilities can fall under more than one definition, for example an ICD that includes a CFS within the site. Moreover, logistics centres have been established throughout Europe, known as Gueterverkehrszentren or GVZ (Germany), Plateformes Multimodales Logistiques (France), Freight Villages (UK), Zonas de Actividades Logisticas (Spain) and Interporti (Italy). These are often attached to an intermodal terminal and, due to the increasing focus on logistics integration (Hesse, 2004), they are often considered as a complete site together, which further confuses the conceptual situation.

Rodrigue et al. (2010) have suggested that the American term 'inland port' be adopted to refer to inland sites, as, like seaports, they encompass the entire site within which numerous activities may be undertaken, of which container handling is only one. Therefore they argue that 'inland terminal' is too restrictive as it would seem to exclude the larger entity, governance structure, etc. In their definition, then, 'inland port' is analogous to 'port' or 'seaport', while a container terminal within the inland port is analogous to the container terminal within the seaport. Likewise with logistics zones, which can be sited within a port or an inland port, and, again with any other entities that may exist within the site, all overseen by some kind of governance body analogous to a port authority. While the use of generic 'inland port' terminology represents an elegant solution for encompassing all kinds of inland nodes, two points need to be made. First, in Europe 'inland port' generally designates an inland waterway port. Second, inland ports in the US are generally far larger than most inland terminals in Europe, some handling several hundred thousand containers annually, therefore supporting large scale warehousing or production districts in the wider area. Thus there are obstacles to using the term 'inland port' to describe an intermodal terminal in Europe that has no water access and may handle fewer than 100,000 containers (in many cases, fewer than 50,000) annually.

The term Inland Clearance (or Container) Depot (ICD), has been a common classification, evincing a particular focus on the ability to provide customs clearance at an inland location (Hayuth, 1980). Similarly, the term 'dry port' has been in use for decades now. It has often been used interchangeably with ICD, and can refer either to an ICD in a landlocked country or to one in a country that has its own seaports (Beresford and Dubey, 1991; Garnwa et al., 2009). The dry port concept in itself is not new, going back at least to UNCTAD (1982), but it has recently returned to prominence in the academic community as a number of journal papers have been published on the topic in the last few years. The 1982 UN definition focused on the dry port as a site to which carriers could issue bills of lading, and this definition was developed in the 1991 UNCTAD document *Handbook for the Management and Operation of Dry Ports*, in which the terms dry port and ICD were used interchangeably (Beresford and Dubey, 1991). Beresford and Dubey (1991) established the basic features of a dry port as well as additional facilities that may be present depending on the local situation. The potential benefits of a dry port were also listed, along with the warning that fiscal incentives and a high level of promotion were often necessary in the early stages of a development.

The primary meaning of both terms (dry port and ICD) is the extension of the bill of lading to an inland destination where customs clearance is performed. Thus the ICD or dry port acts as a gateway 'port' for the inland region. Overuse of the term in recent years has resulted in a multiplicity of understandings; while technically the terms are interchangeable (see Beresford and Dubey, 1991), the term dry port tends to be used instead of ICD to refer to a larger site with many services offered such as storage, containerisation and related logistics activities. It is therefore often used when a site is promoted by public bodies desiring economic benefits for their region through the establishment of such a site. While the transport mode is not an essential part of the definition, a high capacity mode is commonly assumed (most often rail but also inland waterway), as an integral aim of the site is to lower transport costs. This often leads to confusion over whether the term refers to the intermodal terminal or the container processing activities (e.g. customs, storage, etc.), especially as in many cases the two functions are operated by separate companies. More recently, the dry port term has been used in industry as a marketing tool, perhaps to imply that a facility has reached a particular level of sophistication in terms of services offered, such as customs or the presence of Third Party Logistics (3PL) firms within the site and/or an adjoining freight village or similar (see also GVZ in Germany, ZAL in Spain, interporti in Italy).

Why discuss dry ports in the first place? As discussed above, different terminals will have different facilities, but they can be grouped under various headings depending on their primary features, and it has been shown that the terms dry port and ICD have been and continue to be used interchangeably. According to Roso et al. (2009), 'the dry port concept goes beyond the conventional use of rail shuttles for connecting a seaport with its hinterland. Being strategically and consciously implemented jointly by several actors, it also goes beyond the common practice in the transport industry' (p.344). Roso et al. (2009) also state that 'the dry port concept mainly offers seaports the possibility of securing a market in the hinterland' (p.344), and quote the extended gate concept of Van Klink: 'Inland terminals may be considered as 'extended gates' for seaports, through which transport flows can be better controlled and adjusted to match conditions in the port itself' (Van Klink, 2000: p.134).

Dry ports in the original usage were generally developed by inland actors, a requirement emerging from being landlocked or otherwise suffering from poor port access. It thus appears that the Roso et al. (2009) definition may contradict the original definition of a dry port, as the authors prescribe that it is driven by actors from the maritime system; indeed the authors state that a dry port is 'consciously implemented' (p.344) to improve container flows at the seaport. Therefore, as observed by Wilmsmeier et al. (2011), a contradiction can be observed, which is relevant to the discussion within the port regionalisation concept as to whether the inland terminal is an active node imposing on the port or whether the port is consciously involved in developing inland terminals as a strategy of hinterland capture.

Roso et al. (2009) state that: 'A dry port is an inland intermodal terminal directly connected to seaport(s) with high capacity transport mean(s), where customers can leave/pick up their standardised units as if directly to a seaport' (p.341). The key aspect of this definition is the authors' contention that 'for a fully developed dry port concept the seaport or shipping companies control the rail operations' (p.341). One aim of this chapter is to consider to what degree this situation actually obtains in the industry. Are rail operations to sites labelling themselves 'dry ports' run by the seaport or shipping companies?

Additionally, the same dry port definition contends that the seaport and the dry port confront the user with a single interface, with the goal being to provide a smoother operation to users of both the port facility and the hinterland served by the port. The authors further state that their definition includes the 'extended gate' concept, which has been discussed by Van Klink (2000) and more recently by Rodrigue and Notteboom (2009) and Veenstra et al. (2012). Veenstra et al. (2012) defined the extended gate concept: 'seaport terminals should be able to push blocks of containers into the hinterland ... without prior involvement of the shipping company, the shipper/receiver or customs' (p.15), and they claimed that the idea whereby the seaport controls the flow of containers to the inland terminal is an addition to the Roso et al. (2009) dry port concept. However, Roso et al. (2009) do claim that 'for a fully developed dry port concept the seaport or shipping companies control the rail operations' (p.341). This statement may refer to the train haulage rather than the actual decision with regard to container movement, so a potential confusion exists in the overlap between these definitions.

The critical relevance of the dry port concept as developed in Roso et al. (2009) relates to the conscious attempts to secure hinterland markets, and is particularly driven from the seaward side. However, it is not demonstrated in any of the cases assessed by Roso et al. (2009) or Roso and Lumsden (2010) that the seaport or shipping line controls the operations. Therefore, as the extended gate aspect of the Roso et al. (2009) dry port concept is crucial to its stated claim as a consciously used strategy for securing hinterlands, further research is required on this point.

As the first objective of the research is to examine the influence of different inland terminal development strategies in port regionalisation, particularly the active role they play and the role of ports, this issue will be explored through the lens of the consciously-used dry port or extended gate concept. As a contrast, another set of inland terminals without this port relation will be examined in order to learn about the difficulties of integrating or cooperating with ports. If such terminology is to be clarified, the relationship between the port and the inland terminal must be understood in more detail.

Dry port and ICD continue to be used interchangeably, and if the term dry port is to have a distinct identity not synonymous with ICD, it is either to differentiate a site in a landlocked country, or, in the more recent definition, to identify an extended gate of a seaport, where the shuttle is operated by an actor from the seaport as a conscious activity to extend their hinterland. But how many sites would fit this definition? Are the rail links to the European sites that self-identify as dry ports operated by seaports? This issue will need to be tested as site data are compiled.

Developing the Research Factors

Research factors to address the first research question can be derived from the port regionalisation concept in conjunction with the literature just reviewed. Notteboom and Rodrigue (2005) claim that regionalisation is imposed on ports, and that the port is not the main actor, whereas the dry port concept proposed by Roso et al. (2009) claims that dry ports are inland terminals consciously used to capture hinterlands and compete with other ports. Roso et al. (2009) suggest that the rail/barge operations are controlled by the shipping line or other sea actor, rather than a rail operator or inland logistics or transport provider. First, this definition competes with both the regionalisation concept and the earlier dry port concept, which was very much an inland activity, providing administrative services (bill of lading extension, customs, etc.) for a landlocked or otherwise poorly accessible inland region or country. Second, the notion of a sea actor controlling the rail operations seems unusual, and might be covered by the extended gate concept already found in the literature (even if, as shown above, the literature is not always consistent on these definitions and relations with the dry port concept).

Four factors can be used to structure the research, guiding data collection, analysis and comparison:

1. Development process;
2. Relation with ports;
3. Operational issues;
4. Logistics.

These factors can be used to examine different development strategies, different levels of port involvement and integration of rail operations between the port and the inland terminal.

Case Study Selection and Design

The research for this chapter is based on a multiple-case design, therefore replication logic was applied in the case selection. The 11 case studies are divided into two groups. The first set will address the issue of 'consciously implemented' inland terminals, exploring the 'dry port' and 'extended gate' concepts. Therefore each case in this first set has been selected according to a replication logic of these concepts, based on whether they have been developed by a port. However, two types of replication are possible (Yin, 2012: p.146). In this set, the 'dry ports' are direct replicants. The 'extended gate' Venlo is a theoretical replicant, as it was selected to vary from the direct replicant in a predictable way according to theory. This will become clearer in the later discussion.

Six sites were selected for the first group. Four use the term 'dry port' in their name (Coslada, Azuqueca, Muizen, Mouscron/Lille) and the first three of these sites were included in a review of 'dry ports' (Roso and Lumsden, 2010). As only three intermodal terminals developed by ports exist in Spain, and two of these were already selected, it made sense to visit the third (Zaragoza). Finally, the sixth site was the Venlo 'extended gate' system, a concept that shares many similarities with the Roso et al. (2009) 'dry port' concept.

The second set consists of inland terminals not developed by ports. Notteboom and Rodrigue (2005) highlighted a focus on logistics integration, so it was considered important to examine logistics-focused sites. These sites will also follow replication logic by analysing five freight villages in Italy.

The two sets of inland terminals examined in this chapter are listed in Table 4.2.

Table 4.2   List of inland terminal sites for Chapter 4

Country	Set	No. of sites	Site locations
Spain	Port-driven	3	Coslada/Madrid, Azuqueca de Henares, Zaragoza
Belgium (and France)	Port-driven	2	Muizen, Mouscron/Lille
Netherlands	Port-driven	1	Venlo
Italy	Inland-driven	5	Nola, Marcianise, Bologna, Verona, Rivalta Scrivia
Total		11	

The fieldwork for these case studies took place during 2010 and 2011. Each case study was based on a site visit to the inland terminal and interviews with the terminal manager and other representatives, along with analysis of documents obtained at the site and through desk research, supplemented where possible and made more relevant by observations from the site visit.

The case study process was guided by the research factors, based on the literature review. Each case study begins with a brief overview of the freight system in that country, based on desk research. Then a structured narrative of each case based on the four factors is presented, followed by a discussion of pertinent issues. The analysis in this chapter is based on a group of matrices, which show the evidence for each aspect and reveal how inferences were drawn and conclusions reached, through processes of cross-case comparison and pattern matching.

Presenting the Case Studies 1 – Port-Driven Inland Terminals

Spain: Azuqueca, Coslada, Zaragoza [2]

Introduction
Spanish ports are owned by the state, managed by the national body Puertos del Estado and run by port authorities on a landlord model. The only inland terminal in which Puertos del Estado is involved is the Dry Port of Madrid at Coslada, in which the national body collaborated with the four major container ports (see case studies below). While there is no national inland terminal strategy as such, the national body can assist in coordinating initiatives, providing inter-regional coherence to the traditionally regional administration of logistics platform development. As an example, Puertos del Estado is collaborating with the port authorities and regional administrative bodies to consider the potential for inland terminals in Andalucía.

Map 4.1 (below) shows the location of the four major ports in Spain by container throughput. Madrid (the location of Azuqueca and Coslada) and Zaragoza can also be seen.

Container throughput in the west Mediterranean has increased enormously over the last decade (for a discussion of the reasons behind this development see Gouvernal et al., 2005). Table 4.3 (below) shows the container throughput at the top four Spanish ports in 2009. It is interesting to note that Valencia and Algeciras have maintained their traffic while the other two ports have suffered a noticeable fall in throughput.

Bilbao traffic is mostly short sea or feeder from northern range ports in Europe due to its location and Algeciras volumes are mostly transhipment. Valencia and Barcelona are the two major ports for Spanish deep sea cargo, although Valencia does more transhipment than Barcelona. The table also shows the hinterland throughput (i.e. transhipment figures have been subtracted to reveal genuine trade flows).

The geography of Spain means that the hinterland of each port is generally not too far inland so intermodal terminals are not relevant to these flows. The only inland markets of significance are the greater Madrid area (pop. 5–6m) and north-eastern

---

2   This section draws on material from Monios (2011).

Map 4.1   Map of Spain showing location of the four major container ports
Source: Author

Table 4.3   Throughput at top four Spanish ports in 2009

Spain	World	Port	TEU 2009	TEU 2009 (hinterland)	TEU 2008	TEU 2008 (hinterland)
1	27	Valencia	3,653,890	1,829,254	3,602,112	2,023,630
2	34	Algeciras	3,042,759	151,908	3,324,310	159,614
3	58	Barcelona	1,800,213	1,193,917	2,569,550	1,571,962
4	138	Bilbao	443,464	438,818	557,355	543,502

Source: Author, based on Containerisation International (2012); Puertos del Estado (2009)

Spain, which is the primary industrial region in the country. In general, Spain is a net importer, and this is particularly acute in Madrid, so balancing empty container flows is a problem. Catalonia is more balanced because, as the main industrial area, it exports as well as imports. At the Dry Port of Coslada 99 per cent of import containers are loaded, but for exports this figure is only 40 per cent.

Case studies

Table 4.4   Case study table: Azuqueca

Development	Opened in 1995, this was the first such site to be developed in Spain. Planned and developed initially by the port of Barcelona, with private real estate company Gran Europa (which now owns 75 per cent of the site) becoming involved during the process. The remainder is now owned by the ports of Barcelona, Bilbao and Santander. The site was granted a 45 year lease on the land from the local authority, starting in 1994.
Rail operations	Terminal is operated by the majority owner, real estate company Gran Europa. The trains are run by third-party rail operators on a common-user basis. Total TEU has risen from about 2,000 in 2001 up to approximately 25,000 TEU in 2008, before falling to approximately 15,000 TEU in 2009. Of this, roughly 50 per cent is from Barcelona, 40 per cent Bilbao, and 10 per cent Valencia. The services from Valencia and Bilbao to Azuqueca are run by Continental Rail, while TCB runs the rail operations from Barcelona.
Relationship with ports	Integrated through share ownership but little operational involvement. Services to the ports of Barcelona, Valencia, Bilbao.
Logistics	Gran Europa is a real estate company that developed much of the logistics area in the region and then built this terminal to service this demand. But the site itself is just a terminal.
Other comments	Azuqueca also handles bulk traffic such as steel, cereals and cement. 70 per cent of their traffic is containers, 30 per cent bulk.
	Heavily marketed as a 'dry port' for the port of Barcelona.
	The interviewee said that they need a wagonload service to build traffic at the site but they are having difficulty convincing a rail operator to provide one.

Table 4.5   Case study table: Coslada

Development	Opened in 2000, the site was developed jointly by national port body Puertos del Estado and the four major container ports Barcelona, Valencia, Bilbao and Algeciras, with support from Madrid regional government and the local council. Ownership is 10.2 per cent each by Puertos del Estado and the ports of Barcelona, Valencia, Bilbao and Algeciras. The remainder is split between Madrid Regional Government (25 per cent), Entidad Publica Empresarial de Suelo (13.08 per cent) and Coslada Local Council (10.92 per cent). The facility has a 50 year agreement with the local council to use the land.

(continued ...)

(Table 4.5 *concluded*)

Rail operations	After a tender process, the site operation was awarded on a ten-year concession to Conte-Rail which is a private company owned by Dragados (50 per cent), national rail operator RENFE (46 per cent) and Puertos del Estado (4 per cent). However, Continental Rail has been competing for the rail services since 2007. In 2009 the terminal handled 45,000 TEU, down from a high of 60,000 TEU in 2008. Currently the only services are with the port of Valencia.
Relationship with ports	Integrated through share ownership but little operational involvement. However, there is some integration in the sense that the terminal operator is majority owned by the main terminal operator at the port of Valencia (Dragados), which is also the primary source of traffic.
Logistics	There is a logistics platform next door but no direct relation between the two sites.
Other comments	The site was developed jointly by the four major container ports but it now only has traffic with one port and the operator of the terminal is majority owned by the terminal operator at Valencia port. Heavily marketed as a 'dry port' supporting the Spanish port system.

Table 4.6   Case study table: Zaragoza

Development	While the logistics centre ZAL Mercazaragoza is not new, the Terminal Marítima de Zaragoza was only opened in 2009. The terminal site is owned by the company TM Zaragoza, with a shareholding of 56 per cent ZAL Mercazaragoza (the logistics platform), 21 per cent port of Barcelona, 20 per cent from the region of Aragon and the remainder held by local companies.
Rail operations	The terminal is owned and operated by TM Zaragoza. Services are run by third party operators. Throughput in 2009 was 23,864 TEU.
Relationship with ports	The port of Barcelona is integrated through share ownership but has little operational involvement. All traffic is with the port of Barcelona, as the site is within its natural hinterland.
Logistics	The terminal is embedded within and majority owned by the logistics platform.
Other comments	At first the Zaragoza logistics platform was only linked to Barcelona by road, but once the rail corridor to Azuqueca was operational, Zaragoza was a stop on the corridor so it made sense to use it. Originally the distance to Zaragoza was too short to compete against road, but it works now as part of the corridor service. Heavily marketed as a 'dry port' for the port of Barcelona.

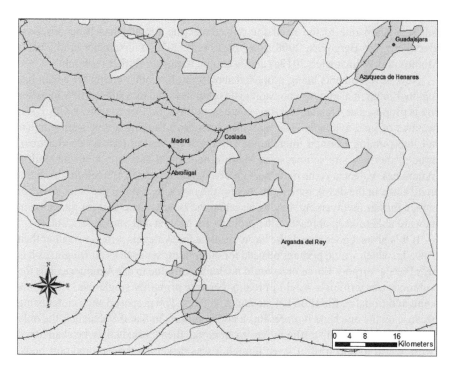

Map 4.2 **Map showing location of current (Coslada, Abroñigal, Azuqueca de Henares) and proposed (Arganda del Rey) rail terminals in the greater Madrid area**
Source: Author

Besides the two intermodal terminals in the Madrid area, as noted above, there is another rail terminal in central Madrid at Abroñigal that acts as a consolidation point for landbridge services between Bilbao and Seville (see Map 4.2). Coslada does not compete for that traffic as it focuses only on rail shuttles directly to the major ports.

The greater Madrid area contains about 5–6m inhabitants and that is the hinterland for the Coslada terminal, but it does overlap with the hinterland of Azuqueca and Abroñigal. The hinterland for Azuqueca includes Madrid, but it is mostly the wider Guadelajara area where there are many distribution centres. In fact, it is the consolidation of cargo to fill a train that can go to both sites that can help to make rail viable.

Because land planning decisions are made at a regional level, getting permission for Coslada with respect to the location of other sites was not a problem because Azuqueca is in another region (Guadelajara, as opposed to Madrid). However, both sites required some additional funding to support the rail connection, therefore limiting the danger of over-saturation of terminal sites. The role of this kind of self-regulation due to the need for rail connection funding has been observed elsewhere (see Bergqvist, 2008; Bergqvist et al., 2010; Wilmsmeier et al., 2011; Monios and Wilmsmeier, 2012a) and is a topic requiring further research.

New developments being proposed show that, like other countries, Spain has regional/municipal bodies who want to develop new logistics sites. A new logistics site is proposed at Arganda del Rey, southeast of Madrid (see Map 4.2) with 1,350 hectares of land available. The plan also includes installing a new semi-circular rail line running from an interchange site north of Madrid (Alcala de Henares), through the new site southeast of Madrid, and round to a site south of Madrid (Aranjuez). Valencia is the main port involved, but the port of Barcelona also has a small stake in the development process for the new site. Even if the latter does not pursue further involvement, having a seat on the Board means that for the moment they are able to keep abreast of the project (see Moglia and Sanguineri, 2003).

If this project goes ahead, the likely result is that Valencia will use it rather than Coslada, which would perhaps be used for other purposes such as air freight, as it is near Barajas airport. Barcelona would no doubt continue to use Azuqueca, thus the common-user terminals would in reality become primarily single-use, with some small additional traffic from Bilbao and Algeciras. The proposed site is interesting because on the one hand it represents a policy failure, in that if Valencia is the only user of Coslada and its traffic moves to Arganda, then Coslada may be abandoned (with regard to port traffic), even though it was driven by the national port body. On the other hand, if all the Coslada traffic is coming from the Dragados terminal, and Dragados holds the controlling share in the concessionaire of Coslada, it may keep the traffic moving through there rather than Arganda (unless Dragados wins the concession for that too), due to the benefits of vertical integration and lower transaction costs.

Discussion

Dragados Marvalsa is the largest container terminal at the port of Valencia, and 90 per cent of the traffic from Valencia to Coslada is from this terminal. Therefore since 80 per cent of the total traffic at Coslada has been from Valencia (100 per cent in 2010), it could be concluded that the 'common-user' terminal is, in reality, a private terminal for Dragados. As was seen above, Dragados owns the controlling share in Conte-Rail, the company operating the terminal. Therefore while officially a publicly-operated facility, there is a degree of vertical integration of a private company.

Similarly, Valencia only provides about 10 per cent of the traffic to Azuqueca. So it is very much a case of Valencia using Coslada and Barcelona using Azuqueca for access to Madrid traffic. Barcelona's involvement in both Coslada and Azuqueca provides security and flexibility, and considering that future capacity at Coslada is limited, Azuqueca gives them longer term security.

Therefore, although much is made of the common-user nature of Spanish terminals, in fact the majority of usage comes from Spain's two large ports, Barcelona and Valencia. Valencia uses Coslada to access Madrid (as a small part of their Madrid traffic, the rest of which goes by road), while Barcelona is able to compete with Valencia by using Azuqueca for Madrid access. Zaragoza is used by Barcelona to access the industrial area in that region, which is in any case within the natural hinterland of Barcelona port. If the future site at Arganda del Rey is developed, this may replace Coslada as Valencia's primary inland node. The effect on competition between the two ports will depend on what inland rates can be offered. It also depends on which shipping lines are calling at which of the two ports. The choice of which inland terminal (Azuqueca or Arganda) is used for Madrid containers will be primarily a result of the port choice (Barcelona or Valencia, respectively).

Since the liberalisation of Spanish rail operations due to an EU directive, a number of private operators have entered the market to compete with the incumbent RENFE. The benefits are now beginning to be seen. In 2007 Continental Rail handled about 10 per cent of the traffic between Valencia and Coslada, but by 2009 it was up to 25 per cent and in 2010 it was closer to 40 per cent. In fact, after two weeks of working with the terminal Continental Rail had captured all of the Maersk traffic from Valencia to Coslada. Rail operations from Coslada to other ports are all through RENFE, but this represents only 10–20 per cent of the total Coslada throughput.

Ports still have problems with the actual rail connections into the port, so infrastructure improvements are required to reduce shunting. At the moment, rail accounts for only a tiny proportion of inland traffic from Spanish ports. In 2008 Valencia handled 69,048 TEU by rail (Fundación Valenciaport, 2010), while Barcelona's rail throughput was 52,562 TEU (in total, including to France) (Port of Barcelona, 2010). This represents just over 3 per cent of hinterland throughput for each port (see Table 4.3).

Reasons for optimism include the upgrading of the rail line from Barcelona to France to European gauge, which is due for completion in 2012. This will allow direct transport without the need to change from Iberian gauge to European gauge. This will help Barcelona in attempts to compete for French cargo, building on its existing rail service to the inland terminal at Lyon. In addition, the new high speed passenger line running from France through Barcelona to Madrid means that the old line is now available for freight traffic, albeit on Iberian gauge. Meanwhile, Valencia has been investing in upgrading rail connections right into the port, as well as developing an IT system that will increase service integration and make rail more efficient and hence attractive to users.

In all three cases, port authorities have formed partnerships with terminal developers and operators. From a port development point of view, it can be seen that the ports are improving their inland access by ensuring terminal facilities in the appropriate locations. Additionally, Barcelona and Valencia are both developing logistics zones within the port perimeter, as well as being involved in inland load centres. Therefore the port authorities at Spain's two largest container ports (excluding transhipment) are pursuing multi-layered regionalisation strategies in partnership with a number of stakeholders.

Muizen, Belgium. Mouscron/Lille, France

Introduction

Freight transport in the Benelux region is understandably shaped by the large northern range ports (see Map 4.3). Therefore inland transport needs to be coordinated with developments taking place in each port, necessitating various business relationships, from joint ventures to vertical integration. The development of the extended gate concept by ECT linking the port of Rotterdam with inland terminals will be discussed in another case study; this section will look at developments in Belgium linking the ports of Antwerp (2009 throughput of 7.3m TEU) and Zeebrugge (2009 throughput of 2.2m TEU) with their hinterlands (Containerisation International, 2012). Sites were visited at Muizen and Mouscron/ Lille, and interviews were conducted with InterFerryBoats and Delcatrans.

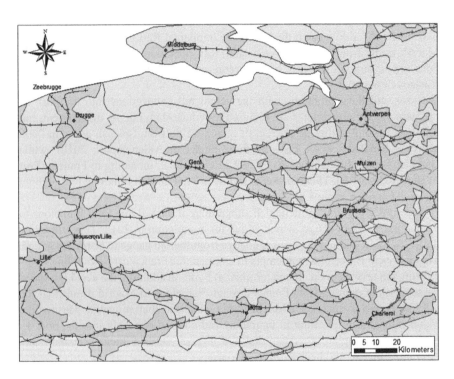

Map 4.3   Map showing the ports of Antwerp and Zeebrugge and the inland terminals at Mouscron/Lille and Muizen
Source: Author

Case studies

Table 4.7   Case study table: Muizen

Development	The site was opened in 1994. All the investment was public. Belgian railways developed the land, built the infrastructure and bought cranes, while IFB (99 per cent owned by the Belgian Railways) paid for the other superstructure. Due to the EU directive, the Belgian Railways was split into infrastructure (Infrabell) and operations (SNCB). SNCB was then split into three subsidiaries. IFB is one of these, focused on containers. Belgian Railways owns the site and IFB leases it from them.
Rail operations	IFB runs the site but handles trains from any company, including the rail operations arm of their own company. To improve efficiency each train has a fixed set of wagons, so they only lift containers on and off. It only runs five services per week, with an estimated throughput of less than 20,000 TEU. The trains handled at the Muizen terminal are company trains, so IFB has nothing to do with the booking, sales, etc. It just handles the full train for a client.
Relationship with ports	Just a normal inland terminal with no specific relationship with any port. The only direct port service is with the port of Zeebrugge. In terms of maritime actors, it is the shipping line that they deal with, rather than the port authority or terminal operator, although they do deal with port terminal operators when setting up the services.
Logistics	There is no direct involvement with logistics; it is just an intermodal terminal.
Other comments	IFB owns four terminals: three in Antwerp plus Muizen. It operates five sites, and participates in some others, working in partnership with TCA (Athus), Delcatrans, (Mouscron/Lille), CDP (Charleroi), LLI -ECE (Liège) and ATO (Antwerp). According to the interviewee, generally if IFB uses a terminal it will have some ownership of it, say 15 per cent, so that it is involved in what goes on there.
	It was noted by the interviewee that, in future, collaborations between terminals and freight villages may be more common.

Table 4.8   Case study table: Mouscron/Lille

Development	Built by the regional government in 2005, it went out of business and was taken over on a 30 year lease by private company Delcatrans. It is run jointly with its main site at Rekkem which is a couple of miles away inside the Belgian border.
Rail operations	Delcatrans operates the two terminals and sub-contracts rail operator IFB to provide traction for their services but it is Delcatrans who deals with the customer. There are 11 weekly services, with 2009 throughput at both sites of 46,000 TEU, all port flows.

(continued ...)

(Table 4.8 *concluded*)

Relationship with ports	It is just a normal inland terminal with no specific relationship with any port. There are services to the ports of Antwerp, Zeebrugge and Rotterdam. Delcatrans deals primarily with the shipping lines, in terms of container throughput, bookings and management.
Logistics	Delcatrans provides logistics services when customers require, and it has some sheds onsite at sister site Rekkem but it is a small business. Most of the customers, but not all, use the full door-to-door service, but take the container offsite. Its core business is trucking; terminals are part of the business, but as the focus is on the door-to-door service, it is difficult to separate them in terms of profitability. Delcatrans also runs a 50,000m^2 logistics platform next door to Rekkem. About 5 per cent of its customers use the logistics platform. The others take the containers to their own warehouses.
Other comments	Initially Delcatrans only ran another site (Rekkem, just inside the Belgian border), but Mouscron/Lille (just over the French border) was not making money and Delcatrans took it over.
	In terms of integration, Delcatrans finds it is easier for the terminal to be independent. It works in partnership with companies like IFB but the interviewee says that they have no need of integration.

Discussion

In these two cases, the relationship between ports and inland terminals is independent but not antagonistic; the large Northern Range ports do not have any direct relationship with these small terminals. The fact that IFB operates a large rail terminal (Mainhub) inside the port of Antwerp means that a degree of operational integration is possible, although the port itself has no direct involvement in the terminal.

These two sites were chosen for study because they are called 'dry ports' (and the third site Rekkem is linked to one of these) and have been quoted in other literature as such (FDT, 2009; Roso and Lumsden, 2010). Therefore the aim was to investigate this claim as a means to scrutinise the use and meaning of this term. It was found that neither of the two sites calling themselves 'dry ports' fits the Roso et al. (2009) dry port vision of an integrated extended gate type operation, and Muizen does not offer customs clearance, meaning that it does not fit the original UN definition of an Inland Clearance Depot. Both are small common-user terminals with little relation to the ports. Interestingly, the interviewee at Delcatrans wondered at any interest in Dry Port Mouscron/Lille as Rekkem is their main site.

These two small intermodal terminals benefit from the subsidised operations of national rail operators. IFB is subsidised by the national rail operator so Muizen is indirectly subsidised, both the terminal operations and the services using it. Similarly, the Delcatrans operation at Mouscron/Lille and Rekkem is indirectly subsidised as it relies on the rail services provided by nationally-subsidised operator IFB.

Venlo, Netherlands

Introduction

Rotterdam is the busiest container port in Europe and the tenth busiest in the world, with a throughput of 9.7m TEU in 2009 (Containerisation International, 2012). Besides calls from the major deep sea lines, the port also offers feeder services to ports all over Europe. Rail, barge and road transport link the port with customers throughout Europe, far beyond its immediate hinterland. However, in common with other large ports, in recent years Rotterdam has experienced congestion problems as a result of its growth, resulting in a number of developments to improve its hinterland access. This case study will focus on the major terminal operator within the port.

As well as conventional rail lines to inland sites, the port of Rotterdam is also the western terminus of the Betuweroute, a multi-billion euro rail line to Germany financed by the Dutch government. This double-tracked, double-stacked electrified line is expected to have a major impact on Rotterdam's access to German customers in the future.

Europe Container Terminals (ECT) began operation in Rotterdam in 1966, and was bought by HPH in 2002. Its throughput in Rotterdam in 2009 was 5.95m TEU. ECT operates three terminals in the port: the lock-restricted ECT City Terminal close to the city of Rotterdam and two deep water terminals in the Maasvlatke development on the North Sea: ECT Delta Terminal and since 2009 the Euromax Terminal. 50 per cent of the new Euromax terminal is owned jointly by four shipping lines: Cosco, K-Line, Yang Ming and Hanjin.

ECT operates a number of inland terminals all linked to the Rotterdam terminals. Located near the German border (see Map 4.4), TCT Venlo is the largest inland terminal in the Netherlands, with a 2009 rail throughput of 115,000 TEU. It offers both rail and barge connections. Over the border in Germany, ECT operates one terminal of the five on a large site at Duisburg, which offers rail and barge transport. The DeCeTe Duisburg site handled 184,000 TEU in 2009. Two smaller sites offer barge transport only. Moerdijk (Netherlands) is close to Rotterdam and acts in some ways as an overspill facility, handling 57,000 TEU in 2009. TCT Belgium (Willebroek) had a 2009 throughput of 76,000 TEU.

All of these terminals are operated in partnership with other companies, and ECT has been endeavouring to develop what it calls the 'extended gate' concept, offering document-free passage of containers from the shipping line through the port to the inland location. Carrier haulage is less than 20 per cent of their throughput, therefore merchant haulage is very important; ECT is developing the concept of 'terminal haulage'. This case study is focused on Venlo, as it is the primary exponent of the extended gate concept. ECT terminals at Rotterdam and Venlo were visited and interviews conducted with site managers.

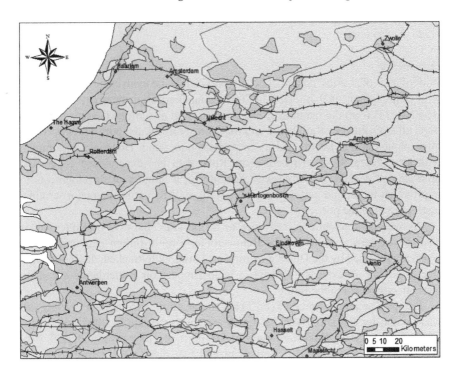

Map 4.4   Map showing locations of the port of Rotterdam and the inland terminal at Venlo
Source: Author

Case study

Table 4.9   Case study table: Venlo

Development	Opened in 1992, with about 15–25 per cent of investment coming from the national government. The regional government was also a shareholder. Otherwise a private initiative and majority owned by ECT.
Rail operations	ECT is the majority owner as well as the operator of the rail terminal. The services are run as a closed loop 'extended gate' system between the port terminal and the inland terminal, managed by ECT with sub-contracted traction. This is in contrast to the Venlo barge, which is independent, and Duisburg, where the local freight forwarder runs the train. 2009 rail throughput of 115,000 TEU.
Relationship with ports	Full integration as the port terminal operator ECT owns the inland terminal at Venlo. 20 trains per week with Rotterdam.
Logistics	ECT has a 50/50 joint venture with Seacon Logistics to operate the adjoining freight village.

Other comments	There was no industry in the area at the time; it has grown with the business, and indeed the terminal was built with warehousing nearby in order to develop this business. This was why they considered it crucial to have a partner on the logistics side. The partnership is successful because, according to Seacon, 'ECT thinks in terms of containers, while we think in terms of the contents'.
	The 'extended gate' system is an interesting example of the port operator taking direct involvement in hinterland flows. ECT does all the booking. They have scheduled services and it is up to ECT to organise the containers on each one. However, ECT can cancel a train if they don't need it. It is all run from Venlo; the operator there will tell the port which containers they need and which train to put them on. Their computer system shows every container on each deep sea vessel, when they are unloaded and where they go. The operator at Venlo can see if the containers he is expecting are not ready to go on the train so he can rearrange schedules where required. Likewise, customers can look up their containers at any time and see exactly where they are. Some customers say they want a specific train, while others just want the container at a certain destination by a certain time and leave it up to ECT.

Discussion

ECT has been using the extended gate concept since around 2007. There have been a number of problems with documentation in terms of using the extended gate concept. First, the operator must be authorised (Authorised Economic Operator) to move containers on behalf of the client. At Venlo they worked on an EU-funded project called INTEGRITY which investigated customs clearance with no checks all the way through the chain. If Seacon is doing the logistics then it is easier for it to plan and manage container movements including documentation because it has knowledge of where the final destination is and other information. Other bottlenecks that need to be overcome are the time taken to book a container on a service, who makes that decision and when.

The development of the Venlo terminal by port terminal operator ECT is a particularly interesting example of a port regionalisation strategy. Operational issues (primarily port congestion) have driven the port actor to develop a hinterland strategy, but they have chosen to integrate fully through acquisition of the intermodal terminal (the logistics platform is a joint venture) rather than through a joint venture or contractual situation as is more common elsewhere. They have developed a model of terminal haulage rather than carrier or merchant haulage, which results in greater efficiency within the closed system.

Notteboom and Rodrigue (2009) noted that the success of such an integrated haulage concept depends on the visibility of cargo in the transport chain. By integrating not only the port and the inland terminal, but also the logistics operation through a joint venture, ECT is able to combine knowledge of the primary and secondary haul requirements, which enables better planning of cargo movements. Veenstra et al. (2012) provided more detail on the information sharing within ECT's extended gate system.

Rodrigue and Notteboom (2009) have discussed how terminals can be used to move beyond push or pull logistics to 'hold' logistics, absorbing time in the supply chain, and that is what is being done at Venlo. In this way, they are able to align the system requirements arising from container management, cargo transportation requirements that drive demand for those containers, and the requirements from the supply chain in which the cargo is embedded. All of these issues then need to be aligned with the vessel management imperative of shipping lines. Indeed, the container movement requirements of shipping lines can raise difficulties for inland container management, when containers used for merchant haulage must be returned immediately to the port for repositioning by the shipping line.

Another interesting aspect of the Venlo development is that it straddles two classifications. It is a load centre serving a major source of transport demand, and the trimodal terminal is surrounded by logistics and supply chain facilities. However, the operational integration between the port terminal and the inland terminal makes the site function as a satellite terminal, operationally integrated with the container yard in the port. This kind of operation is generally more likely to be found in close proximity to the port, functioning as an overspill facility to provide an extension to operations, rather than being linked to shippers far inland.

Therefore this case study demonstrates several interesting developments in inland terminal operations and functions. However, as noted above, it still requires much work on the part of stakeholders to overcome legal and practical obstacles to achieve the full potential of the site, and then to extend the system further through ECT's other inland terminals.

Like many European countries, there is some evidence of optimism bias in government support of inland terminals in the Netherlands. The interviewees complained about terminals that have been built with government subsidy but end up not being used. According to the interviewees, in the 1990s inland terminals (the Netherlands) and GVZs/freight villages (Germany) were springing up around this part of Europe, as it is the heart of the industrial zone, but not all of them had a solid enough market to survive the current economic climate. So the interviewees raised some questions about the potential misalignment of public subsidy with market and operational realities.

Presenting the Case Studies 2 – Inland-Driven Inland Terminals

Italy: Marcianise, Nola, Bologna, Verona, Rivalta Scrivia

Introduction
Table 4.10 lists the ports in Italy with the highest container throughput, and the top five are shown in Map 4.5.

Italian ports are run on the landlord model; they are publicly owned and the terminals are privately operated on a concession basis. There is no national body to coordinate them as there is in Spain. According to one interviewee, port

Table 4.10　　Top ten Italian ports by container throughput (2010)

Port	Coast	TEU
Gioia Tauro	South	2,851,261
Genoa	West	1,758,858
La Spezia	West	1,285,455
Livorno	West	635,270
Taranto	South	581,936
Cagliari	Sardinia	576,092
Naples	West	532,432
Venice	East	393,913
Trieste	East	281,629
Salerno	West	274,940

Source: Author, based on Containerisation International (2012)

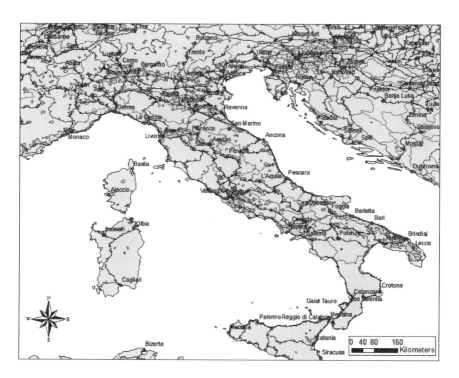

Map 4.5　　Map showing top five Italian container ports
Source: Author

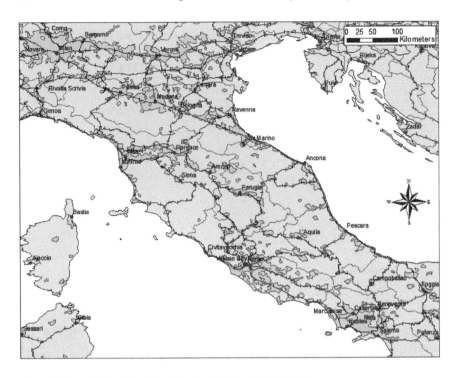

Map 4.6   **Map showing the five Italian freight villages**
          Source: Author

authorities have to give most of the revenues to the national government, leaving little for reinvestment in the port; the port authority must then either obtain private investment or else ask the national government for money. Therefore there may be a misalignment of strategy between national and local scales, but that is a separate topic of research, beyond the scope of these inland terminal case studies.

Italy exhibits a distinctive model of interporti or freight villages, which are large logistics platforms with attached intermodal terminals. The definitive aspect of this model is that the focus is firmly on logistics as much as transport. Almost all of these sites have had public involvement at some point in their development, and many retain public-private ownership models. Unlike most of the case studies in this research, the intermodal terminal has not been built as an independent site but has been built into a logistics platform (usually from the outset but in some cases added later), and remains a key but less significant part of the overall business at the site.

There are currently 24 members of the Italian interporti association (UIR: Unione Interporti Riuniti). Most of these sites are in the north of the country, where the majority of industry and production centres are located, close to the heart of Europe. While interporti were developed through a variety of mechanisms (see case studies), they have gradually been brought into a national planning strategy. The National Transport Master Plan (PGT) of 1986 identified first and second level interporti, and the next version in 1990 devolved responsibility to the regional level. National law 240/90 was important because it officially recognised interporti in a national network, making them eligible for national funding. In order to be considered a freight village under this law and thus be eligible for funding, the site must include an intermodal terminal. All interviewees noted the importance of this law, although there was disagreement on whether the money was distributed fairly and indeed how much each had received. According to the official figures, a total of 533m euros was spent by the national government on all the interporti between 1992 and 2003 (UIR, 2009).

The five freight villages visited for this research are shown in Map 4.6.

Case studies

Table 4.11   Case study table: Marcianise

Development	Opened in 1999. Developed with mostly private money although did receive some federal grants (although amount is disputed). Owned and operated by private company Interporto Sud Europa.
Rail operations	Interporto Sud Europa owns Rail Italia (which runs the terminal) and Rail Services Logistics (which operates the trains and deals with the clients). As well as the large terminal within the site, there is a very large marshalling yard (the biggest in Italy) operated by national operator RFI/Trenitalia just outside the terminal. They will handle trains by any operator but currently the trains are run by their own operating company Rail Services Logistics. They are trying to have an integrated service between their own or partner terminals but it is difficult to compete with the nationally owned operator Trenitalia due to their government subsidy. Precise details on current services were difficult to obtain, but despite a very large intermodal terminal they run only a handful of services, with container throughput estimated at below 10,000 TEU.
Relationship with ports	There are currently no services to ports, although there was a service with Naples in the past. The interviewee noted that they have had difficulty establishing good relations with ports. The interviewee felt that the ports do not cooperate and will only do so if they are very congested and have no choice. When the interporto did work with the port, it was the terminal operator not the port authority with whom they worked.
Logistics	As with all the interporti, logistics is the main focus; this site concentrates on industrial and manufacturing clients.
Other comments	The site is not yet complete so there is a large amount of land still to be developed.
	The interviewee said that the shunting yard outside the site was built with EU funding with the intention of linking to the port of Gioia Tauro, but this traffic did not develop.

Table 4.12   Case study table: Nola

Development	In 1989 the National General Transport Plan identified the need for a freight village in the Campania region. Interporto Campania is a private company that was awarded the right from the region to build and operate the site, which was opened in 1997. The rail terminal was opened in 2006.
Rail operations	The terminal is operated by Terminal Intermodal Nola, which is owned 60 per cent by Interporto Campania and 40 per cent by Galozzi (the operator of Salerno port). They also started their own train company Interporto Servizi Cargo in 2009 which both provides traction and deals with customers. That is the only company currently running trains there but others can if they want. 2010 throughput was 25,250 units or approximately just over 40,000 TEU. There is also a very large shunting yard, just outside the terminal but within the overall site boundary, owned and operated by national operator RFI/ Trenitalia.
Relationship with ports	A daily container service runs to the port of Naples. The terminal operator is building a closer integrated relationship, and the operator of the rail terminal at the port of Naples is owned jointly by the Naples port authority, Interporto Campania and national operator Trenitalia so there is some vertical integration there.
Logistics	As well as the interporto, there is a large wholesale distribution centre built in 1986 (Centro Ingrosso Sviluppo – CIS). Many customers use both the CIS and the interporto. As with all the interporti, logistics is the main focus; this site concentrates on retail and wholesale clients. It is mostly in-house logistics provided here rather than by 3PLs like at Marcianise.
Other comments	Much of the rail freight is for customers at the site, but the interviewee said that probably the majority goes outside the site. The terminal handles a mixture of containers and swap bodies on their services with Bologna, Milan and Verona, whereas their Naples service is 98 per cent containers.
	It is interesting that the only port service is with the port of Naples, and the rail terminal at the port is partly owned by the Nola interporto. However, the rail terminal at Nola is partly owned by the port of Salerno, with no investment from the port of Naples.

Table 4.13   Case study table: Bologna

Development	Opened in 1980. It was mostly public money at first to set up the site, while private investors came later. Ownership of the company is 35 per cent the municipality of Bologna, 18 per cent the province of Bologna, 6 per cent the Bologna chamber of commerce, 23 per cent of shares are held by banks, 16.5 per cent by private companies and 1.5 per cent by Trenitalia.

Rail operations	Interporto Bologna owns the freight village but national rail operator Trenitalia owns the two large intermodal terminals, and they are operated by Terminali Italia (the terminal operating arm of RFI/Trenitalia). Interporto Bologna is planning a new intermodal terminal that it will own and run. This is to overcome problems it is having with Trenitalia Cargo. These problems were described as twofold: as operator, Trenitalia is cutting services, and as the infrastructure provider, it is not investing. Trains to the site are run by third-party operators. In 2010, the terminal handled 190,000 TEU, with mostly inland origins and destinations.
Relationship with ports	Just a normal inland terminal with no specific relationship with any port. They have direct services with the ports of La Spezia, Livorno, Ravenna and Ancona, and via Piacenza to Rotterdam and Zeebrugge. The interporto has agreements to collaborate with ports. When they have done this, it has been with the port authority rather than the terminal.
Logistics	As with all the interporti, logistics is the main focus. The majority of users of the intermodal terminal are outside the freight village. However, the aim of the interporto is to get more customers to use rail.
Other comments	While the aim of the interporto is to get more customers to use rail, it is not a requirement for site customers that they must use it if they locate here.

Table 4.14   Case study table: Verona

Development	Consorzio ZAI is a fully public company that owns and operates the site (three shareholders: town, province, chamber of commerce). The interporto was built in the late 1960s, with the rail terminal built in 1977. Consorzio ZAI is like a port authority; it does not run anything in the site, but just manages it. The aim is not to maximise profit but to develop logistics infrastructure in the region, therefore profit is re-invested in the company. It builds the infrastructure and rents the warehouses to clients.
Rail operations	The two main intermodal terminals are owned by Quadrante Europa Terminal (owned 50 per cent by Consorzio ZAI, 50 per cent by RFI/Trenitalia) and operated by Terminali Italia (part of RFI/Trenitalia). There are also two small ones. Trains are run by third-party operators. The vast majority of traffic is swap bodies rather than containers, so it is almost all internal European traffic. There is very little port traffic. In 2010 the site handled 327,433 units (equating to 480,017 TEU by their calculations).
Relationship with ports	Just a normal inland terminal with no specific relationship with any port. Most of their traffic is intra-European; the only port with a direct service is La Spezia.

(*continued ...*)

(Table 4.14 *concluded*)

Logistics	As with all the interporti, logistics is the main focus; but they are like a port authority and don't deal with users directly.
Other comments	About 80 per cent of their intermodal traffic goes to customers outside the site.

Table 4.15   Case study table: Rivalta Scrivia

Development	Developed initially in 1963 as a 'dry port' for the port of Genoa, so there was a specific focus on clearing customs inland as the port was congested, although it was not possible to get more detail on what was meant by that designation. This is in contrast to the other interporti. Interporto Rivalta Scrivia is fully privately owned (68 per cent Fagioli Finance, 22 per cent F21 logistics, 8 per cent by other private companies and 2 per cent by the region). The rail terminal opened in 2006 and is owned by Rivalta Terminal Europa, which is itself owned 47.87 per cent by Interporto Rivalta Scrivia, 47.87 per cent by the Gavio Group, and the remainder is owned by public partners: the Piemonte region, the port authority of Savona, the province of Alessandria and the township of Tortona.
Rail operations	Rivalta Terminal Europa owns and operates the terminal, and is also involved in booking slots and selling the train services to customers. The rail operators are just traction providers. About 90 per cent of their rail business is from ports, especially the port of Genoa, 75km away, and of that, about 90 per cent is from Voltri Terminal Europa. Maritime containers are their main equipment. This is different to the others which handle mainly swap bodies. Throughput in 2010 was approximately 150,000 TEU.
Relationship with ports	Direct services to the ports of Genoa, Savona and La Spezia. Currently managed through contracts but they are discussing a potentially closer collaboration with the ports. The interviewee said that the port is considering becoming a shareholder in the site. The terminal deals with the port terminal (Voltri) rather than the port authority, as well as shipping lines. The method of service development is by way of contracts rather than full integration. The most important client of the terminal is the shipping lines.
Logistics	As with all the interporti, logistics is the main focus, but the difference here is that Rivalta Scrivia is a logistics operator so they deal directly with the clients rather than 3PLs doing it. This is the reason, according to the interviewee, that most of its rail traffic is for the site customers whereas for other sites most of the rail throughput is for customers outside the site. The terminal mostly attracts the business of big companies, mainly shipping lines (therefore carrier haulage): e.g. CMA CGM, Maersk, MSC.

Other comments	There are problems with congestion at the port of Genoa therefore rail can compete over this distance. In addition, the port is surrounded by mountains, which makes road haulage less attractive.
	A new intermodal terminal is being built, with the goal of an estimated capacity of 500,000 containers annually (double the current capacity). Unlike the current intermodal terminal which is inside the interporto border, the new larger terminal is just outside.

Discussion

Freight villages in Italy can get public funding due to being designated an official freight village (and meeting the conditions) under the 1990 law, but it was difficult to get a clear answer from any of the interviewees on the specifics of the funding award process. Many sites were built long before the law came into being so they did not receive money in their start-up period. The amount of government money depends on meeting certain criteria so each project has to justify its request to the Ministry of Transport. However, although it is a national law, the sites are developed at regional and local level. The sites are naturally jealous of each other's receipts of funding from the national level. Most sites were publicly planned (the most common model is a public-private partnership [PPP]), but unusually, Verona is fully public. There are currently 24 freight villages in the association, and the government is in the process of drafting a report to update the 1990 law.

An opportunistic interview was obtained with the operator of a distribution company for hardware/DIY/homewares, who imports 95 per cent of his product through the port of Salerno (from Asia, mostly China). He distributes in southern and central Italy, solely by road. 'Rail does not exist in Italy', he said. He said that road regulations are not adhered to in Italy, resulting in many overweight trucks and long driving hours. He fills his containers completely rather than using pallets because that would leave empty space in the container. Using pallets would be more convenient but would cost more due to this wasted space. There is no problem taking these over-filled containers on the road because he says that hauliers do not adhere to the regulations.

One interviewee at a freight village expressed deep scepticism about the potential for intermodal transport to develop in Italy. There are only a few shipping lines competing for the sea leg, but this interviewee suggested that there are 200,000 transport companies in southern Europe alone, with far more in Italy than in countries such as France and Germany. Therefore the system is much more fragmented. Moreover, a lack of regulatory oversight, illegal driving hours and overloaded weights makes it very difficult for rail to compete with road. Even within companies, departments are very fragmented so it is hard to put together a new transport model. The potential for Gioia Tauro to be a significant gateway for Italian traffic is constrained because the shipping lines do not want it to be, according to one interviewee.

One interviewee said that Europe has many small sites, but not a proper linked system of infrastructure with major nodes, therefore what is needed is a system of major hubs linked by regular shuttles like a road system. Rail used to be a single national body with knowledge of all the network, brownfield sites, old rail heads, etc. so they were better at utilising assets cheaply and making it work. Now this knowledge has been lost due to fragmentation and the break up of organisations and institutional knowledge.

The Rivalta Scrivia freight village in the hinterland of the port of Genoa has high port traffic and it is even working towards a potential trial of an extended gate concept (for more detailed discussion see Caballini and Gattorna, 2009). The terminal has a good relationship with the port, unlike many other freight villages in Italy, but this is because the port needs the inland node due to its congestion issues. This is not the case with other ports in Italy. The port of Naples has problems with congestion and long dwell times, and Iannone (2012) showed that it can actually be cheaper to send the container by rail to an inland node even at relatively short distance, due to the saving of dwell time charges. However, problems of fragmented transport operators and the inability to build cooperation between organisations have prevented these services from prospering.

A key point to note when discussing freight villages as integrated sites is that, first, the general model is for the intermodal terminal to be operated by a separate company, and second, the terminals are common-user facilities, with some shippers located within the freight village and some not. Indeed, the majority of the rail traffic at large freight villages such as Bologna and Verona is actually for companies outside the site. Rivalta Scrivia proves an exception to this rule, as the operators of the site work directly with the shippers located there rather than through 3PLs.

Analysing the Case Studies

All the data were reviewed and relevant information was entered into the individual tables that were constructed for each case. The meta-matrices presented below have collated the key data from each case, facilitating cross-case analysis. The data collection, primarily through interviews but supplemented with document analysis and observation, was guided by the pre-determined factors. Thus, while the analysis proceeds by induction, a strict grounded theory approach is not followed here.

Tables 4.16 and 4.17 set out the thematic meta-matrices, with the relevant data noted against each factor.

The meta-matrix has been divided in two, one for port-driven terminals and one for inland-driven. These labels were used based on the discussion in the literature review, where the two classes were established. However, it was found in the analysis that, not only is it difficult to claim with certainty which organisation took the lead in a development process, in many cases even the 'port-driven' sites were not actually led by the port actor. However, the terminology will be retained during the discussion, as that was how the data collection and analysis were structured.

Table 4.16 Key features of the inland terminals (port-driven)

Country	Location	Owned	Operated	Customs onsite	Logistics in same site	Driver (organisation)	Driver (public/ private)	Relation with port	Method of integration with port	Info sharing	Who controls rail operations
Spain	Azuqueca	Mixed	Private	Yes	No	Port authority/ private investor	Mixed	Med	Partial investment	Med	Rail operator
Spain	Coslada	Public	Private	Yes	No	Port authorities/ region/ municipality	Public	Med	Partial investment	Med	Rail operator
Spain	Zaragoza	Mixed	Private	Yes	Yes	Port authority/ private investor/region	Mixed	Med	Partial investment	Med	Rail operator
NL	Venlo	Private	Private	Yes	Yes	Port terminal operator	Private	High	Ownership	High	Inland terminal

Table 4.17 Key features of the inland terminals (inland-driven)

Country	Location	Owned	Operated	Customs onsite	Logistics in same site	Driver (organisation)	Driver (public/ private)	Relation with port	Method of integration with port	Info sharing	Who controls rail operations
Belgium	Muizen	Public	Public	No	No	Rail operator	Public	Low	None	Low	Rail operator
France	Mouscron/ Lille	Private	Private	Yes	No	Region	Public	Low	None	Low	Inland terminal
Italy	Marcianise	Private	Private	Yes	Yes	Private investor	Private	Low	None	Low	Rail operator
Italy	Nola	Private	Private	Yes	Yes	Region/private investor	Mixed	Med	Joint ownership of rail terminal in port	Low	Rail operator
Italy	Bologna	Mixed	Mixed	Yes	Yes	Municipality/ region/ private investor	Mixed	Low	None	Low	Rail operator
Italy	Verona	Public	Public	Yes	Yes	Municipality/ region	Public	Low	None	Low	Rail operator
Italy	Rivalta Scrivia	Private	Mixed	Yes	Yes	Private investor	Private	Med	Contracts	Med	Rail operator

Some cases that were selected under the port-driven category were in fact not developed by ports. Dry Port Muizen and Dry Port Mouscron/Lille were both selected because they call themselves 'dry ports' and Muizen was included in the Roso and Lumsden (2010) review of 'dry ports'. Findings from the research revealed that these sites were actually developed by inland actors, either rail operators or regions, so they have been put into the second matrix along with the freight villages. Yin (2012) discussed how to deal with cross-case synthesis processes when the actual cases turn out to be different from what was thought during the screening process. It is possible to analyse the cases as theoretical replicants ('predicted to have different experiences, but with conceptually consistent explanations, p.146) rather than direct replicants ('predicted to follow courses of events similar enough that they repeat or replicate each other's experience in a conceptual, not literal, sense', p.146).

Beginning with the port-driven terminals, results show that Venlo was an example of a terminal driven by a private port terminal operator, while all three terminals in Spain were examples of terminals driven primarily by public port authorities. The Venlo case has been arguably the most successful, perhaps because the port terminal operator is directly involved in the operations at the inland terminal, whereas the port authorities in the Spanish cases are not. Not only is the terminal operator ECT involved in the intermodal terminal, but a 50 per cent joint venture in the logistics platform means greater information sharing is possible. There is therefore a possibility of developing different conceptual models according to whether the port actor is the authority or a terminal operator. Ports can use a variety of mechanisms to coordinate the hinterland transport chain and thus reduce transaction costs (de Langen and Chouly, 2004; Van der Horst and de Langen, 2008; Van der Horst and Van der Lugt, 2009), but full integration is rare, and thus stands out as an innovative development. Venlo represents an attractive hinterland access strategy, but many institutional, operational and legal difficulties prevent comparable developments elsewhere (Veenstra et al., 2012).

In Spain, public port authorities are involved in inland terminal development, yet despite their heavy marketing of this fact, in none of these sites do the port authorities own a majority shareholding or direct the operations. By contrast, ECT is a private port terminal operator actively integrated with the inland site. While in all the cases above a port actor is involved to some degree, the results reflect the difficulty for a port authority (and, to a lesser extent, a port terminal operator) to exert influence beyond the port's perimeter (de Langen, 2008; Moglia and Sanguineri, 2003). The underlying theme is that port actors want improved inland access to further their business aims but they can rarely be said to be driving these developments. In most cases the port actor is a partner in someone else's plan. Similarly, with the exception of Venlo, all inland terminals in the above sample are independent from the port.

Frequency counts are not relevant in this kind of analysis as statistical inferences may not be drawn from a theoretically-driven sample. However, Tables 4.16 and 4.17 reveal the expected result that the port-driven inland terminals in this sample have a higher level of port involvement. Table 4.16 shows that in the Venlo case the port terminal operator owns the terminal rather than a minority shareholding; it therefore has a close relationship with the terminal and information is shared via their container management system, as discussed in the case study. Furthermore, the rail operations are managed via a sub-contracted traction provider, which means that a closed system operates between the port terminal and the inland terminal. In none of the Spanish cases does this kind of operation exist. The inland terminals are independent from the port and the rail operations are on a common-user competitive basis, just as with any intermodal terminal; they are not integrated with the port in any meaningful way beyond conventional practice.

While the cases have shown that ports can actively develop inland terminals, the analysis has revealed that the reality in practice is often overstated. Of the six sites identified from the literature as port-driven, two (Muizen and Mouscron/Lille) were not developed by ports at all, while three (the Spanish sites) have minority port investment, but are otherwise run as conventional inland intermodal terminals. Only Venlo can be considered a genuine port-driven site or a genuinely distinct regionalisation strategy. Therefore questions are raised about the ability or actuality of ports integrating with inland terminals.

By comparing Table 4.16 and Table 4.17, it can be seen that there is a difference between terminals developed by port actors and those developed by inland actors. In all of the port-driven cases, the port actor does have some investment in the inland terminal, although it differs between each one. Likewise, the operational relations are higher in the port-driven group, as would be expected. However, the inland-driven group have no investment from the port and low operational cooperation. A likely explanation is that if a port actor is involved from the beginning in developing an inland terminal, then motivation exists to maintain a good relationship to achieve the operational efficiencies or other benefits that were the motivation for the initial investment. Whereas if the port actor is not involved from the beginning, then inland actors struggle to obtain port involvement later. Further explanation of port motivations for inland investment and cooperation and the limitations on their ability to do so would require additional research on ports, which is beyond the scope of this chapter.

Table 4.18 groups the cases by the main driver, in order to follow a pattern-matching procedure to look for conceptual groups.

It is difficult to draw definitive conclusions from the data in Table 4.18, as more cases are required. However, it can be seen that port-driven terminals take different forms, whether they are driven by port authorities or port terminal operators. It appears from the cases above that port terminal operators are better placed institutionally to forge a closer relationship with the inland site due to their ability to integrate operationally, while a public port authority cannot. This underlies the requirement for information sharing through logistics operations. This is difficult, hence most port-driven terminals may be expected to have a

Table 4.18 Four models of inland terminal development

Cases	Driver			
	Port authority	Port terminal operator	Rail operator	Public body
Cases	*Coslada; Azuqueca; Zaragoza*	*Venlo*	*Muizen*	*Mouscron/Lille; Nola; Marcianise; Bologna; Verona; Rivalta Scrivia*
Government role in development	Some regional and local government investment	There was some national government subsidy in the initial development of the site and some share ownership by the regional government	The government was not directly involved but the rail operator who developed the site is nationally owned	Most were developed by the regional or local government, but some had majority or total private investment
Government role now	Regional and local government retains some investment	Regional government retains some investment	Indirect subsidy of national rail operator	M/L: none Italy: all the sites developed by government retain a percentage of local/regional government investment, but the level varies. Also, since the 1990 federal law, federal funds are available for all, even privately-developed sites
Port role in development	Port authority investment and initial impetus	Port operator was main investor in the inland terminal	None	None except Rivalta Scrivia, which was developed as a 'dry port' from the beginning
Relations with port	Good relations and info sharing but no direct integration	Port operator owns and operates the inland terminal and manages the rail services (with sub-contracted traction)	Just standard operational relations	Low at all except developing good relations at Rivalta Scrivia, and some integration of rail operations between Nola and the port of Naples

(continued …)

(Table 4.18 concluded)

Cases	Port authority	Port terminal operator	Rail operator	Public body
			Driver	
	Coslada; Azuqueca; Zaragoza	Venlo	Muizen	Mouscron/Lille; Nola; Marcianise; Bologna; Verona; Rivalta Scrivia
Rail operations	Third party, no specific integration or cooperation	Managed directly by use of sub-contracted traction, closed loop, extended gate	Terminal owner/operator is one arm of the national rail operator and it handles trains of any company, including the rail operating arm of its own company	M/L: terminal owner/operator handles trains of any company, including its own (sub-contracted traction) Italy: terminal operations operated by different companies at each site, mostly either owned by the interporto or the national operator RFI/Trenitalia Terminals will handle trains from any company, although in some cases the main operator is owned (at least partly) by the interporto
Logistics role	Coslada: freight village next door Azuqueca: surrounded by logistics/ warehousing of individual companies Zaragoza: situated within logistics platform	Port terminal operator has a 50 per cent JV in the adjoining freight village	None	M/L: low. The operator provides logistics as well as transport services but it is only a small company Italy: high. All sites are primarily logistics platforms
Motivation for building site	Hinterland capture	Hinterland capture as well as improving port operations to reduce congestion. Also the port of Rotterdam requires a certain level of containers to go by rail	An operational decision, to serve customers in that region	M/L: current owner did not develop the site but it is speculated that it was to support local/regional businesses Italy: support local/ regional businesses. Rail use is more important now as they must have a rail terminal to meet the criteria of the 1990 law allowing federal funds

purely transport focus, i.e. to get the container out of the port as soon as possible rather than any true operational integration.

The role of all intermodal terminals in this context is the transport function of changing mode between road and rail. However, as load centres serving zones of production or consumption, which is the point of bundling container flows on high capacity links, logistics is important. Locating a terminal within a logistics site is one way to improve the feasibility of the services, by removing the last mile. The Italian sites were all built to this model, although the take-up of the rail services has varied. In most cases they have struggled to attract port traffic.

All rail terminals at the Italian freight villages were much larger than the sites in the port-driven category, sometimes with more than one terminal within the site, including over 10 tracks in some instances and also with lengths up to 800m per track. Therefore these sites are designed for very large rail traffic, making them quite different to the smaller sites in the port-driven category. The majority of the traffic is domestic European traffic, however. The three northern freight villages with regular intra-European traffic had very high intermodal throughput (including both containers and swap bodies), whereas the two southern Italian freight villages, despite large rail terminals, had very low traffic.

It is not necessarily that inland-driven terminals are logistics-focused, but that port-driven terminals are less likely to be so, due to institutional and operational reasons. It may be that logistics tends to be more important for those sites developed by public bodies, as the aim is to support businesses, but further research is required. In the Italian cases, the majority of users of the intermodal terminal were not in fact site tenants, which was an unexpected finding. More detailed analysis of this point would be valuable.

Results from field work in Italy showed that the freight village concept is good for logistics, but has had very little success integrating with ports. Indeed, even getting rail traffic at all is not easy due to the road-dominated and fragmented Italian logistics system (Evangelista and Morvillo, 2000), and some freight villages have very large intermodal terminals with very low rail traffic. 'Rail does not exist in Italy' was the comment of one shipper. Yet this is not the entire story, as was already illustrated in the comparison above between the ports of Genoa and Naples in respect of their relations with inland terminals.

All of the inland-driven terminals were developed predominantly by the public sector. Most were developed by the regional government, although most of the Italian freight villages have had some private investment. In Muizen, it was the publicly-owned rail operator that initially developed the site, although now the terminal operating arm of the company is operationally separate from the rail operating part, due to the EU directive on liberalisation. It is interesting that, for all terminals that had direct government involvement, the scale of government was predominantly regional rather than national.

A potential conflict can be identified between two broad conceptual groupings: on one hand a port-driven, operationally focused, potential satellite terminal/ extended gate concept and on the other, a public-sector-driven load centre concept.

Wilmsmeier et al. (2010, 2011) borrowed from the terminology of industrial organisation (i.e. forward and backward integration) to introduce a conceptual approach to inland terminal development by providing a directional focus absent in the priority corridor model of Taaffe et al. (1963). In this model, the authors contrasted Inside-Out development (land-driven e.g. rail operators or public organisations) with Outside-In development (sea-driven e.g. port authorities, terminal operators). This distinction is a shorthand way of identifying potential strategy conflicts between actors with different motivations, and will be discussed in Chapter 8.

A lack of integration between port and inland systems can be observed in the above case analysis, and even antagonism suggested by many inland terminal interviewees. While more case studies of alternative practice are required, the case studies in this chapter show that even in instances where port-inland integration is desired, the actuality is rare and, except for Venlo, only focused on the transportation function (i.e. moving the container inland), whereas the logistics and supply chain functions are more the interest of inland actors. Even the success of the Venlo example requires further legal and practical barriers to be overcome before its potential can be reached.

Thus the first three findings from this analysis are that ports can develop inland terminals, there are different ways of doing it (i.e. port authority or port terminal operator) and, finally, that differences have been observed between port-driven and inland-driven sites. The fourth finding from this chapter is the role of the 'dry port' concept.

To begin with, two of the 'dry ports' (Muizen and Moscron/Lille) were not developed by ports and retain no port investment or operational involvement. The other two 'dry ports' (Azuqueca and Coslada), while they were at least partially driven by port investment, are common-user terminals with competitive rail operations, so there is no extended gate system or port actor controlling the rail operations. Table 4.17 shows that one 'dry port' (Muizen) does not even have customs, so it does not even fit the original dry port definition which is that it should provide inland clearance. While the use of the 'dry port' terminology by these sites is inconsistent, the 'extended gate' operation at Venlo is the only 'consciously implemented' terminal in this sample. Indeed, it can be considered the only example of a true port regionalisation strategy of all the cases analysed in this chapter.

Dry Port of Coslada/Madrid, Dry Port Azuqueca, Dry Port Muizen and Dry Port Mouscon/Lille all use the term 'dry port' but they function differently. IFB (the operator of Muizen) runs terminals but it just handles the trains of other companies (including trains of a separate part of their parent company IFB Intermodal). At the Delcatrans terminals (Mouscron/Lille and Rekkem), a rail operator is sub-contracted to provide the traction but Delcatrans does all the bookings and container management. So these two terminal types are a contrast, but what they have in common is that no port actor is involved in any of their operations.

What is even more curious is that Delcatrans runs two sites in conjunction: LAR Rekkem (on the Belgian side of the border) and Dryport Mouscron/Lille, just on the French side. The two sites are only a few miles apart and are run jointly.

Dryport Mouscron/Lille was set up by the regional government and went out of business before being taken over by Delcatrans. It is called a 'dry port' because of its initial naming, but both sites are the same – simply small intermodal terminals with a couple of rail tracks and some warehousing nearby. Indeed, the interviewee expressed curiosity at interest in Mouscron/Lille as it is the smaller of the two sites and the main Delcatrans office is at Rekkem.

Neither the two Spanish sites, nor Muizen or Mouscron/Lille would be considered 'dry ports' using the Roso et al. (2009) definition. None of these sites have reached the ideal situation in which 'the seaport or shipping companies control the rail operations' (Roso et al.: p.341). In only the Spanish cases can the port be said to have 'consciously implemented' these sites. Yet these four sites use the term in their site names, which is why three of them (Coslada, Azuqueca, Muizen) were included in a review of 'dry ports' (Roso and Lumsden, 2010).

The distinctive aspect of the Roso et al. (2009) definition seems to be the close link between the port and the inland site. The matrix in Table 4.19 presents one way of categorising such developments, applying a pattern-matching process.

Table 4.19   Matrix showing different levels of integration in port-inland systems

Does the port actor manage the inland haulage, i.e. container slots, sales, etc.	Port involvement in the terminal	
	Yes	No
No	Coslada/ Madrid Azuqueca Zaragoza	Muizen Mouscron/Lille Nola Marcianise Bologna Verona Rivalta Scrivia
Yes	Venlo	

Note: this table is an expansion of a table previously used in Monios (2011)

Both Spain and Venlo give examples where the port is involved. The difference is that in Spain it is the port authority, whereas with Venlo it is the terminal operator ECT. Furthermore, in the case of ECT, the port terminal is directly involved in the operations, unlike in Spain where it is just a minority shareholder. Consequently if one asserts that the dry port concept involves an integrated service offering, it is exemplified more by ECT's extended gate concept than by those sites using the dry port terminology. ECT is developing the concept of 'terminal haulage' as opposed to the already understood notions of merchant or carrier haulage. Similarly, the port of Valencia has been working on increasing integration with

Coslada by developing a port community system to share information in a single unified system, but at this stage it is purely an information management system.

The extended gate system between Rotterdam and Venlo is perhaps the best example of the Roso et al. (2009) 'dry port' definition, which envisages a combination of an Inland Clearance Depot (ICD) with a freight village, incorporating extended gate integration with the port operations. Interestingly, ECT does not use this terminology, preferring instead the 'extended gate' term. Rodrigue et al. (2010) classified Venlo's extended gate operation as a satellite terminal, as it is fully integrated with the port terminal stack management and can therefore be used as an extension of the port yard or a kind of overspill system. However, unlike a simple overspill or extended yard function, Venlo is also a load centre serving a large hinterland market.

The extended gate or terminal haulage concept (as opposed to carrier or merchant haulage) has been discussed by Van der Horst and de Langen (2008), Notteboom and Rodrigue (2009) and Veenstra et al. (2012). Veenstra et al. (2012) defined the extended gate concept thus: 'seaport terminals should be able to push blocks of containers into the hinterland ... without prior involvement of the shipping company, the shipper/receiver or customs' (p.15), and they suggested that the idea whereby the seaport controls the flow of containers to the inland terminal is an addition to the Roso et al. (2009) dry port concept. However, Roso et al. (2009) do claim that 'for a fully developed dry port concept the seaport or shipping companies control the rail operations' (p.341). This statement may refer to the train haulage rather than the actual decision with regard to container movement, so a potential confusion exists in the overlap between these definitions.

It was noted in the literature review that the earliest dry port definition, as well as being synonymous with ICD, made some reference to landlocked countries using the terminal as a maritime access point, primarily for customs rather than necessarily for intermodal connection. A new definition was proposed by Roso et al. (2009), suggesting that the port actor controls the rail operations, resulting in a combination of an inland clearance depot with adjoining freight village and extended gate functionality. In almost all cases in this chapter, this definition does not apply. The sole example of this level of integration is Venlo, which does not currently use the 'dry port' term.

It is therefore suggested that the 'extended gate' terminology be retained to refer to a specific concept of integrated container flow management between the port and the inland site. The concept of 'terminal haulage' (as opposed to carrier or merchant haulage) represents a new stage of integration that could hold significant potential if technical and operational obstacles can be overcome. The extended gate terminal haulage concept can also be related to a move from push to pull logistics strategies or even 'hold logistics', as outlined by Rodrigue and Notteboom (2009) in their concept of supply chain terminalisation, whereby inland terminals are actively used to manage inventory flows.

By contrast, for most interchange sites (especially in Europe), 'intermodal terminal' or 'inland terminal' may be better terms to describe the common denominator linking the majority of sites; functional analyses can then focus on the activities of each node, for example whether they involve customs clearance, value-added services or overspill functions for a port. Therefore functional distinctions prove themselves to be of greater utility than overall terms.

The inland port terminology adopted in the United States is fairly flexible, focusing primarily on the intermodal terminal itself, but offering the potential to include a multiplicity of sites, including freight villages and any or all transport modes. This was discussed in the literature review, whereby Rodrigue et al. (2010) had suggested the use of this term to cover all inland freight nodes. Reservations were raised there on the applicability of this term to small European sites. It is a potentially useful term, but in usage it tends to designate large gateway sites. As the average intermodal terminal in the USA is much larger than in Europe, this does not pose a classification problem.

The Italian sites represent a distinctive model of interporti, which fit a clear model based on a national transport strategy. They also align closely with other terms as noted in the literature review: freight village, ZAL, GVZ and logistics platform. There is no current academic debate on the taxonomy of these sites. The key distinction is that their primary identification is as a logistics platform providing warehousing and all associated services to a range of users. The transport mode is secondary to this function. In Italy, all officially-recognised interporti offer road and rail transport, while at other logistics platforms in Italy, as with those in other countries, rail may not always be present. The difficulty for many intermodal terminals in Europe achieving economies of scale leads to the conclusion that aligning intermodal terminals with such freight village concepts should be an obvious way to address this problem. However, as was seen with Nola and Marcianise, developing intermodal traffic remains difficult, even with excellent facilities.

Rodrigue et al. (2010) drew useful distinctions between the functions of different sites, classifying them as satellite terminals, transmodal centres and load centres. This functional approach is similar to the close, mid-range and distant dry port model presented by Roso et al. (2009) and the later seaport-based, city-based and border-based model proposed by Beresford et al. (2012). This kind of functional approach, based on the usage of each node, has more utility than overall terms such as 'dry port' or 'inland port'. It allows a research agenda to be developed along the lines of the purpose and usage of these nodes in the transport chains that they shape. It also focuses more clearly on the transport operations of the node, as represented in the actual terminal or interchange point, and, in addition, is more closely aligned with the infrastructure requirements and investment in the site, particularly in terms of planning and public involvement. The 'co-location' of warehousing, logistics, etc. at or near the site tends to result from a number of decisions from individual private firms, therefore attempting to include a potential multiplicity of freight villages or logistics clusters within the umbrella of the terminal concept makes classification and taxonomy development increasingly difficult. Therefore the cases in this chapter serve to strengthen the conceptual distinction proposed by Rodrigue et al. (2010), whereby transport and supply chain functions are categorised under separate taxonomies. The governance relationships between the transport and logistics functions of a site will be explored in Chapter 7.

Conclusion

Before discussing the relevance for port regionalisation, some findings can be drawn from the specific case analyses. Inland terminals have experienced difficulties attracting port flows unless a port actor has been involved from the beginning. The only case that can be called a true success in attracting port flows is Venlo, where the port terminal operator is directly involved in the operation of the inland terminal. A hypothesis can therefore be proposed that inland terminals developed on the basis of intermodal flows with ports can only be successful if a close operational relationship, if not full integration, with the port terminal operator is established from the outset. However, this possibility requires further research. Most interviewees noted that they experienced difficulties establishing good relationships with port actors, suggesting that port and inland systems remain separate in most instances, although again this point requires further research. Suggestions of integration with ports and leaving the container at an inland terminal 'as if directly to a seaport' (Roso et al., 2009: p.341) can only be possible if a number of difficult obstacles have been overcome. Therefore integrated transport chains with lower transaction costs and increased efficiency are not yet the norm and may not be for some time. The empirical analyses in Chapters 5 and 6 address these issues in more depth.

Results in this chapter showed that ports can actively develop inland terminals, and differences exist between those developed by port authorities and those developed by port terminal operators. Examples were provided of both port authorities and port terminal operators directly investing in inland terminals. In the cases studied, the most successful model was the port terminal operator, and a suggested explanation was that this is because it was directly involved in operating the inland terminal and the rail shuttles. The port authority is rarely in a position to do this, thus limiting the potential for successful inland terminal developments by port authorities. A 50 per cent joint venture in the logistics platform at Venlo provides the possibility of greater visibility of flows, which enables further efficiencies.

It was found that there are differences between inland terminals that are developed by ports with a conscious aim of hinterland capture and those that are not developed in this way. Those developed by land actors tend to focus on domestic traffic and even when they attempt to develop port traffic they have difficulties engaging with the port. This division suggests that the integration of maritime and land systems implied by the port regionalisation concept faces several challenges. The difficulties highlighted in the case studies represent good reasons why this integration may not be forthcoming in many instances; not just actual integration, which, as Notteboom and Rodrigue (2005) note, is difficult for ports, but even operational cooperation. Inland transport and logistics systems remain fragmented, particularly intermodal transport which is the focus of this book. Until these operational issues are resolved, even high levels of cooperation may be prevented, much less any chance of integration. These operational issues will be considered in detail in the following chapter.

Finally from a conceptual perspective, it was found that the sites in this sample calling themselves dry ports do not fit the 'dry port' concept proposed by Roso et al. (2009). The only site that fits the definition (Venlo) is already known as an 'extended gate' concept. The importance of separating transport and logistics functions when classifying inland terminals, an approach drawn from the literature (Rodrigue et al., 2010), has been strengthened by the findings in this research, as it was shown that even in sites where an intermodal terminal is embedded within a larger logistics platform, the two functions remain separate.

More cases are required to validate these findings, but the details of how a case works in practice can raise issues about the extent to which port regionalisation can actually happen and what is required for it to happen. The cases elucidate good reasons why ports may not be controlling or capturing hinterlands through the strategies of integration that the port regionalisation concept suggests.

Chapter 5
Case Study (UK): Intermodal Logistics

Introduction

This chapter analyses a case study of intermodal logistics in the UK. Large retailers are the primary drivers of intermodal transport in the UK, and they are explored in the context of their relationships with rail operators and 3PLs. The crucial role of 3PLs in providing a necessary link between the supply chain requirements of the retailer and the operational requirements of rail operators is demonstrated. Results show that inland logistics markets exhibit spatial centralisation and a lack of integration between market players, and the efficiency of rail freight services is challenged by the need to combine port and domestic movements which have different product, route and equipment characteristics.

A Review of the Literature on Intermodal Logistics

Logistics and Intermodal Transport

As discussed in chapter 2, Notteboom and Rodrigue (2005) state that 'regionalisation results from logistics decisions and subsequent actions of shippers and third-party logistics providers' (p.306), and that 'the transition towards the port regionalisation phase is a gradual and market-driven process, imposed on ports, that mirrors the increased focus of market players on logistics integration' (p.301). They go on to note that 'logistics integration ... requires responses and the formulation of strategies concerning inland freight circulation. The responses to these challenges go beyond the traditional perspectives centred on the port itself' (p.302). Before ports can integrate inland, the specific characteristics of inland freight circulation must be examined, in particular the 'logistics decisions and subsequent actions of shippers and third-party logistics providers' (p.306).

Menzter et al. (2004: p.607) defined logistics management as 'a within-firm function that has cross-function and cross-firm ... aspects to it'. It involves the management of demand, in particular through information management, as well as directing supply, involving strategic distribution decisions relating to transport, inventory levels and location for storage and intermediate processing. Many of these logistics decisions, as part of a wider focus on supply chain management, exert considerable impact on transport flows, operational requirements and location decisions (Hesse, 2004). Additionally, the notion of transport solely as a derived demand has been challenged and reformulated as an integrated demand (Hesse and Rodrigue, 2004; Rodrigue, 2006; Panayides, 2006). As such, the relationship between goods flows and spatial development is complicated by networks of nodes and corridors that may not perform their key functions adequately, potentially constrained not just by physical infrastructure deficits but a lack of connectivity or an inability to fit into wider networks. The focus of this research is on the use of rail transport; a firm's decision to shift to this mode can be driven by many factors, such as external pressures (e.g. fuel price, legislation, customer pressure) or logistics strategy (e.g. central warehouse or distributed network, private fleet or 3PL) (Eng-Larsson and Kohn, 2012). Yet, according to some authors, the role of transport in logistics and the broader field of supply chain management has been under researched (Mason et al., 2007).

As a result of Woodburn's (2003) investigation into the relationship between supply chain structure and the potential for modal shift to rail, he concluded that 'for rail freight to become a much more serious competitor to road haulage would require considerable restructuring of either the whole logistical operations of companies within supply chains or far-reaching changes to the capabilities of the rail industry to cope with the demands placed upon it' (p.244). Eng-Larsson and Kohn (2012) established that when making a decision to use intermodal transport, the convenience of the purchase was more important than the price. Taking an operational perspective, they found that other supply chain decisions had to be made to incorporate intermodal transport, such as increasing inventory, extended delivery windows, and improvements in planning and ordering due to less flexible departure times of intermodal transport. However, in each case the additional cost was offset by the savings in transport costs from the new transport mode. They even postulated that some firms may be paying unnecessarily for a higher degree of transport quality (e.g. flexibility, reliability, frequency) than they actually need, thus they could reduce these requirements while still achieving their supply chain objectives.

Several challenges have been identified in the literature regarding modal shift from road to rail. Customers require low transit time, reliability, flexibility and safety from damage, and it has been suggested that the industry perception is that intermodal transport cannot provide these (RHA, 2007). Access has been reduced as the UK rail industry has seen a major decline in wagonload services over the last few decades. Better information for potential shippers is also required regarding train services, timetables and wagon capacity. Due to a lack of marketing and information availability, rail is often not visible to prospective customers. There is also a suggestion that the true cost of rail movements should be more visible to users. MDS Transmodal (2002: p.49) found that 'there are no published rates for rail freight charges and rail freight users have only a poor understanding of their suppliers' cost structures as there are dominant operators in the market and little on-rail competition'.

Break-even distances for intermodal transport are generally considered to be in the region of 500km (Van Klink and Van den Berg, 1998), yet only 22 per cent of freight journeys in Europe are above this distance (Bärthel and Woxenius, 2004). However, the nominal competitive haulage length of 500km can be shortened by other factors such as removing the road leg at one end and the existence of regular unitised demand to ensure high rates of utilisation. Granting exceptions to current length limits for road vehicles feeding intermodal terminals has also been shown to decrease the cost of the pre- and post-haul and thus increase the competitiveness of intermodal transport (Bergqvist and Behrends, 2011). In the UK, break-even estimates for a route that requires no road haulage have been estimated as low as around 90km. With a road haul at one end only, the figure is roughly 200km, and if both pre- and end-hauls are required, the distance is approximately 450km (MDS Transmodal, 2002). The economic difficulties of developing rail shuttles have been addressed in the literature from different perspectives, including transport cost analysis (e.g. Van Schijndel and Dinwoodie, 2000; Ballis and Golias, 2002; Arnold et al., 2004; Racunica and Wynter, 2005; Janic, 2007; Kreutzberger, 2008; Limbourg and Jourquin, 2009; Kim and Wee, 2011) and the importance of aligning cargo requirements with intermodal service requirements (e.g. Woodburn, 2003; Woodburn, 2011; Eng-Larsson and Kohn, 2012).

Assessing the suitability of a particular product flow for intermodal transport requires an analysis of such considerations as the lead time and size of orders, the value and the physical characteristics of the product. Challenges to intermodal transport include distance, lack of flexibility, lead time for service development and the role of the last mile (Slack and Vogt, 2007). The high fixed costs of rail operators and the requirement to consolidate flows on key routes make profitable service development difficult. The complexity of setting up a rail service is another barrier to intermodal growth as well as a barrier to market entry for new operators (Slack and Vogt, 2007).

Cooperation is required to achieve economies of scale on key routes, but research has found industry reluctant to pursue such a strategy (Van der Horst and de Langen, 2008). Moreover, shippers are reluctant to commit to a service unless it is already well developed (Van Schijndel and Dinwoodie, 2000). This feeling underpins a severe inertia in the industry. Runhaar and van der Heijden (2005) found that even a 50 per cent increase in transport costs over ten years would not be sufficient to make producers relocate their production or distribution facilities. This inertia can in some ways be considered a bigger obstacle than infrastructure problems, and needs a restructuring of the transport chain in order to change transport requirements.

The Spatial Development of the Retail Sector

Several trends have been observed in the literature, such as the centralisation and relocation of plants and distribution centres, a reduction in the supplier base and a consolidation of the carrier base (Lemoine and Skjoett-Larsen, 2004; Abrahamsson and Brege, 1997; O'Laughlin et al., 1993). Market power has been concentrated among a few large retailers due to mergers and acquisitions (Burt and Sparks, 2003). Supply chains are being reconfigured around rationalisation of transport requirements and new distribution strategies and hub locations.

(Lemoine and Skjoett-Larsen, 2004). Distribution centres are being optimised and new purpose-built facilities are appearing. This ongoing process of rationalisation means that trying to embed them in intermodal chains is difficult.

The retail sector has evolved over the last few decades from a system whereby suppliers delivered directly to stores to the introduction of distribution centres (DCs) in the 1970s and 80s to the arrival in the 1990s of primary consolidation centres (PCCs) (Fernie et al., 2000). Lead times and inventories were reduced substantially as part of impressive efficiency advances over this period.

Large retailers in the UK drive distribution patterns to a significant degree, not just transport movements but location of facilities. McKinnon (2009: p.295) found that 'since 2004, roughly 60% of the demand for large DCs has come from retailers'. Large retailers have reduced the number of their DCs while increasing the size and efficiency of those that remain. This system of fewer, larger DCs means not simply greater centralisation but potentially greater miles travelled, yet it also raises the possibility for increased use of intermodal transport due to the ability to consolidate flows on key routes. Food and grocery companies currently contribute one in four of all truck miles travelled in the UK (IGD, 2012). Direct container train services from UK ports to the Midlands have grown over the last decade while direct services from UK ports to Scotland (i.e. Coatbridge) have fallen (Woodburn, 2007). This finding represents the integration of Scottish trade flows into UK-wide distribution networks centred on key sites in the Midlands and to a lesser extent north England.

The spatial development just described was built around the motorway network, but some discussion of locating import-focused distribution centres at ports has taken place in recent years (Mangan et al., 2008; Pettit and Beresford, 2009; Monios and Wilmsmeier, 2012b). In the context of the recent increase in research on the importance of inland terminals, perhaps a renewed focus on the potential of the port as a logistics hub is warranted. Revisiting the potential of port-based versus inland-based logistics can even be viewed as another name for optimising the primary and secondary legs of the supply chain, challenging the inertia of supply chains that were constructed in different contexts. Major supermarket retailers Tesco and Asda have both located large general merchandise import centres at the port of Teesport.

A centralised UK inland network, developed when industrial and retail inputs were primarily UK-sourced, may have some drivers to decentralise, focusing on the processing of imports arriving at coastal ports. From a port's perspective, this allows them not only to secure cargo throughput, but to earn additional revenue from these activities on their land (Pettit and Beresford, 2009). Import containers are offloaded from incoming vessels, shunted to the port warehouse, stripped, and the empty then returned for repositioning. This load will then be reconfigured for inland movement. This movement may be direct from the port-based DC to the final store, thus removing the inland DC from the chain.

From an operational perspective, increasing use of ICT has allowed more accurate forecasting and more responsive ordering (thus a move from push to pull replenishment), which in turn required a more tightly optimised spatial distribution of facilities, as well as an integration and optimisation of primary and secondary networks. Thus some retailers work closely with hauliers to optimise their distribution (e.g. reducing empty running or reducing inventory holding requirements) or collaborate with suppliers to optimise product flows (e.g. forecasting, planning and ordering). These relationships are becoming increasingly important in intermodal transport as working rail into a supply chain requires much closer relationships and greater knowledge sharing between partners to solve operational issues. These spatial and operational evolutions have resulted in an increasing integration of operations, ranging from progressively efficient use of backhauling to the implementation of factory gate pricing (Mason et al., 2007; Potter et al., 2007).

When the retailer purchases the supplied goods 'ex works' or 'from the factory gate', rather than paying the supplier to deliver the goods to the DC, it gives the retailer greater control over the primary distribution leg. This can result in greater transport efficiencies (even more by the use of primary consolidation centres), reducing empty and part-full running in both directions, and also improve planning efficiency and responsiveness through greater visibility of flows. It has been estimated that total distance travelled can be reduced by around 25 per cent through the use of a primary consolidation centre (Potter et al., 2007), while another study found that total cost reduction from the use of FGP can reach approximately 8 per cent (le Blanc et al., 2006).

Potter et al. (2007) discussed how FGP strengthens the negotiating position of the retailer. It gives the retailer greater power over the supplier, as the supplier loses the opportunity to cross-subsidise a lower product price with a higher transport price. It also enables the retailer to operate or sub-contract the primary distribution as a single large business rather than negotiating with many small hauliers working for single suppliers. An additional result from this practice is to make it more difficult for smaller transport operators to enter the market. Therefore large 3PLs could benefit while independent hauliers would lose. Burt and Sparks (2003) also noted that non-FGP retailers will find their supply prices increasing because for the supplier the transport cost per unit for non-FGP flows will increase as volume is removed from this stream to the FGP retailer's distribution network. Similarly, Towill (2005) showed how the category management paradigm allows supermarkets to streamline their supplier base and increase their margins.

The increasingly complex transport requirements resulting from these developments led to greater use of 3PLs by retailers. The need for more frequent, smaller deliveries from suppliers to reduce inventories also encouraged the use of primary consolidation centres (Smith and Sparks, 2004; Fernie and McKinnon, 2003). Distribution facilities evolved from single-product warehouses to composite environments housing ambient, chilled and fresh produce, all scanned in and out using barcodes that were integrated within the IT system used for forecasting, planning and ordering. In 1992 Tesco replaced 26 single-temperature warehouses with nine composite sites (Fernie et al., 2000). The use of composite warehouses and trucks as part of increasingly sophisticated temperature-controlled supply chains has doubled the shelf life of some items (Smith and Sparks, 2009). Reverse logistics also became more important, with packaging and other recycling travelling back up the chain from the store to the DC.

Collaboration with competitors is another significant topic in the literature. Successful retail intermodal logistics includes the retailers themselves as well as rail operators and 3PLs. Schmoltzi and Wallenburg (2011) showed that while almost 60 per cent of 3PLs in their study operated at least one horizontal partnership, the failure rate was below 19 per cent, against an average failure rate for horizontal collaborations in many industries ranging from 50 per cent to 70 per cent. This is an encouraging result, but to what extent retailers can put aside their intense rivalries and collaborate on transport, particularly filling trains, will be a key determinant in the future potential of intermodal transport. Schmoltzi and Wallenburg (2011) also found that, while horizontal collaboration might be thought to be based on cost reduction, the primary motivations revealed in their study were service quality improvement and market share enhancement. In a similar study, Hingley et al. (2011) found that cost efficiencies from horizontal collaboration were less important to grocery retailers than retaining control of their supply chain.

Developing the Research Factors

From the literature review, it can be seen that the market has consolidated through mergers and acquisitions, the spatial distribution of distribution centres has been rationalised and centralised, and retailers have achieved greater control of both primary and secondary distribution. Collaboration is less well understood, as it appears to occur in some instances but not others. Likewise, it is known that intermediaries such as 3PLs are playing a greater role in the process, intervening between large shippers and transport providers, but it is not clear precisely how they are involved in the shift towards intermodal transport, relating back to the differing strategies of integration and collaboration between shippers, 3PLs and rail operators. Operational transport issues include the need to consolidate flows despite a decline in wagonload traffic, the inertia in the industry and issues relating to lead times and process planning.

Four factors can thus be used to structure the research, guiding data collection, analysis and comparison:

1. Spatial development of the market;
2. Operational rail issues;
3. Strategies of integration and collaboration;
4. The government role in developing intermodal transport.

Case Study Selection and Design

Woodburn (2012) identified large retailers as the main drivers of intermodal traffic growth in the UK and called for interviews with them to understand the reasons behind the observed growth, in addition to calling for an examination of the role of port-hinterland flows in relation to domestic intermodal routes. Many studies have looked at the supply chain evolution of retailers and to a lesser extent the logistics and transport implications of these changes. These studies have tended to be published in the supply chain, logistics and business management fields. On the other hand, intermodal transport has been studied in detail in the transportation literature, with a tendency to focus on operational aspects to improve efficiency and thus reduce cost, making intermodal transport more attractive to prospective users. This chapter will bring these two fields together in an investigation into large retailers as the key drivers of intermodal transport in the UK. Lessons learned from their success may contribute to an understanding of modal shift in other market or geographical contexts.

The umbrella concept of port regionalisation implies that ports can capture and control hinterlands through intermodal corridors, which are supposedly more monolithic due to strategies of logistics integration on behalf of market players. However, these market players make their transport decisions based on a variety of issues that are not explained in the port regionalisation discussion. It was considered that this successful and growing intermodal market would provide good data on what features of inland freight circulation need to be addressed in the port regionalisation concept if ports are to integrate with this market through intermodal corridors and terminals. Therefore this case was chosen to examine these aspects.

The fieldwork for this case study took place during 2011 and 2012. Interviews were conducted with general freight stakeholders in the intermodal transport industry as well as a number of large retailers, 3PLs and rail operators. The data collection process was guided by the research factors, based on the literature review. The analysis in this chapter is based on a matrix, which breaks down the four factors into 23 sub-factors, clarifying how inferences were drawn and conclusions reached.

The data were analysed as a single case study on intermodal transport use by retailers, with multiple embedded units of analysis (being each organisation: the retailers, rail operators and 3PLs). It could have been structured with each retailer as a case study if the goal were simply to compare their experiences of modal shift. But then there would be repetition and overlap regarding the use of rail and 3PLs by each retailer, and the perspectives of the 3PLs and rail operators would have been pushed to the background. Successful retail intermodal logistics involves numerous actors, therefore the perspectives of all three major stakeholder groups must be considered together.

Presenting the Case Study

Major Retailers in the UK

UK retailers account for almost 6 per cent of UK GDP and employ approximately 3 million people (Forum for the Future, 2007; Jones et al., 2008). The grocery retail sector is extremely concentrated; nearly 83 per cent of the retail market of grocery trade in the UK is controlled by five retailers: Tesco (31 per cent), Asda (17 per cent), Sainsbury (16 per cent), Morrison (12 per cent) and the Co-operative (7 per cent) (Scottish Government, 2012).

Map 5.1 shows the location of the distribution centres of the five major grocery retailers in the UK (PCCs are not shown). The centralisation in the Midlands is clear, as is the lack of coverage in north England, north Scotland and Wales.

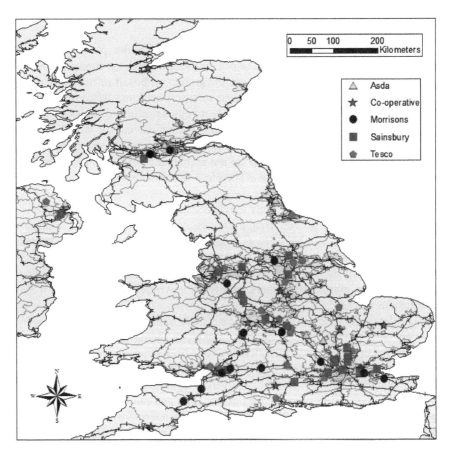

Map 5.1   Map showing the distribution centres of major supermarkets
        Source: Author, based on data obtained from retailer websites

While the sector continues to evolve, retail logistics in the UK has largely reached the maturity stage. The fast-moving, highly-responsive system observed in Britain does, however, exhibit some differences from other countries. Fernie et al. (2010) explained that in countries such as the USA, Germany and France, property costs are lower, therefore retailers use this space to hold higher inventory, the costs of which are offset by the discounts obtained from large orders. This stock is forecast and then pushed to stores, discounted if necessary, and customers tend to buy larger amounts of this discounted product.

Fernie et al. (2010) stated that Tesco is the market leader in logistics, as with sales. Asda is upgrading its systems in line with parent company Wal-Mart, Morrisons is still processing its Safeway merger, while Sainsbury has spent the first decade of the twenty-first century regaining market share after a difficult period caused by an unsuccessful adoption of a new technological supply chain management system.

In this research the focus is on grocery retailers rather than other retail sectors such as fashion, and a wholesaler has been included as a contrast. The major grocery retailers, however, also sell a significant proportion of non-food lines, which is of particular relevance when considering port flows. Grocery retail is generally broken down into a number of distinct categories, including grocery/ambient (which can be further split into fast and slow moving), fresh, chilled, frozen, convenience, non-food and direct. Other areas such as recycling (reverse flows) can also be considered. Each one of these has different operational requirements, which affect their suitability for intermodal transport; for example, non-food lines tend to have more stock keeping units (SKUs) than food because of product differentiation.

Overview of Rail Freight in the UK

Network Rail, a nominally-private but fully government-owned company, owns and operates the track infrastructure, while intermodal terminals are owned or leased by private operators. Private rail operators compete to run services on the shared-user track. The four primary rail freight operators in the UK are DB Schenker (formerly EWS), Freightliner, Direct Rail Services (DRS) and First GBRf. The other key actors are third-party logistics service providers that charter trains from these operators, including John G Russell, WH Malcolm and Eddie Stobart.

Intermodal transport first developed in Britain as a consequence of the maritime container revolution in the 1960s. Distribution centres centralised in the Midlands became key cargo generators and attractors (see Chapter 2). As any port could service the same hinterland, maritime container flows shifted from local ports to the large south-eastern ports.[1] Port-hinterland container services have continued to grow in recent years, with a 56 per cent increase in the number of

---

1 Other factors also influenced this concentration – see Hoare (1986); Overman and Winters (2005); Pettit and Beresford (2008).

these services between 1998 and 2011 (Woodburn, 2012). Domestic intermodal traffic was marginal in earlier years and was utilised primarily for industrial products. Over the last decade this market has grown, primarily due to retail flows, with Asda first using rail in 2003 and Tesco following in 2006. These container flows are moved primarily on the Anglo-Scottish corridor (between terminals in the Midlands and central Scotland). These developments were to some extent subsidised by successful use of government funding for intermodal terminals (Woodburn, 2007). The flows are comprised primarily of northbound secondary distribution of picked ambient grocery loads from retail DCs in the Midlands, back loaded with southbound flows from Scottish suppliers, such as soft drinks and spring water (FTA, 2012). The concentration of DCs and intermodal terminals in the Midlands and in central Scotland, with suitable distance between them (see Map 5.1), has produced a high-density Anglo-Scottish corridor with a short 'last mile' between DC and intermodal terminal at either end. The question is to what extent rail operators are matching the service characteristics of road haulage, to which retailers are accustomed.

The majority of rail freight in the UK has always been bulk traffic, until containers overtook coal for the first time in 2010 (see Figure 5.1).

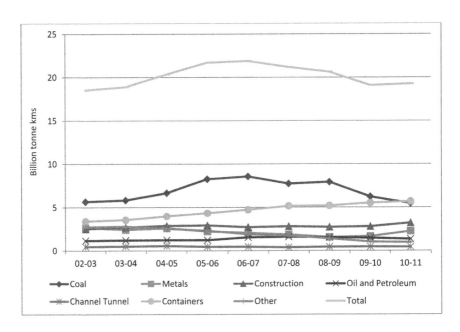

Figure 5.1   Rail freight moved by sector
Source: Author, based on ORR (2012)

Bulk cargo is effectively captive as not only is it suited to rail carriage but origins and destinations tend to have handling equipment built into the sites. This case study is focused solely on intermodal flows. The majority of inland container rail traffic in the UK is between Scotland and England; most of this traffic is anchored at the English end of the route at DIRFT Daventry. Currently handling around 175,000 lifts per year, it is the busiest inland intermodal terminal in the UK.

Table 5.1 lists all current intermodal rail services on the Anglo-Scottish route (not only those used by retailers), split into two categories: ex port (direct service between a port and a Scottish terminal) and domestic (between inland terminals in England and Scotland).[2]

Table 5.1   List of current intermodal rail services running on the Anglo-Scottish route

Type	Service	Traction	Management	Frequency per week
Ex port	Felixstowe – Coatbridge	Freightliner	Freightliner	5
	Southampton – Coatbridge	Freightliner	Freightliner	5
	Tilbury – Coatbridge	Freightliner	Freightliner	5
	Liverpool – Coatbridge	Freightliner	Freightliner	5
Anglo-Scottish	Tilbury-Barking-Daventry-Coatbridge	DRS	JG Russell	2 daily × 5/6
	Daventry – Mossend (DB Schenker)	DB Schenker	Stobart	6
	Daventry – Mossend (PD Stirling)	DRS	WH Malcolm	5
	Daventry – Grangemouth	DRS	WH Malcolm	6/7
	Hams Hall – Mossend	DB Schenker	DB Schenker	5

Source: Author, based on interviews and document analysis

These intermodal services are mostly common-user. The ex port services are booked for the most part by shipping lines as carrier haulage is dominant in the UK for port flows, but smaller users can also book space on these trains directly with Freightliner or through a 3PL or freight forwarder. The other flows are generally managed by 3PLs serving a variety of customers. The largest sector utilising these trains is the retail sector (Woodburn, 2012).

---

2   Intra-England and intra-Scotland services are not shown.

Use of Intermodal Transport by Retailers

Woodburn (2003: p.245) noted that 'it is notoriously difficult to identify specific rail freight users and volumes from public sources, particularly in the non-bulk sectors'. Similarly, a large consultation with stakeholders in the freight industry found that 'information on freight movements is not currently available at an adequate level of detail to reflect the underlying supply chain characteristics' (WSP et al., 2006: p.vii). For this research a list of users was compiled from the interview data and desk research (primarily FTA, 2012). Results are presented in Table 5.2.

Table 5.2   Retailer use of intermodal transport

Retailer	Route	Rail operator	Management
Tesco	Anglo-Scottish	DB Schenker	Stobart
Tesco	Scotland to north	DRS	Stobart
Tesco	Daventry-Tilbury	DRS	Stobart
Tesco	Daventry-Magor	DRS	Stobart
Sainsbury	Anglo-Scottish	DRS	JG Russell
Morrison	Anglo-Scottish	DRS	JG Russell
Waitrose	Anglo-Scottish	DRS	WH Malcolm
M&S (DHL)	Anglo-Scottish	DRS	WH Malcolm
Asda	Anglo-Scottish	DRS	WH Malcolm
Asda	Scotland to north	DRS	DRS
Co-operative	Anglo-Scottish	DRS	WH Malcolm
Costco	Anglo-Scottish	DRS	JG Russell

The majority of companies from Tables 5.1 and 5.2 have been interviewed for this research and will be discussed in this chapter, but in this section all retailers will be considered in order to provide a complete overview of the UK retail intermodal network.

Only one retailer is large enough to move significant flows by rail. Tesco has four dedicated services, matching secondary distribution of picked loads with inbound primary flows, filled out with other materials such as packed-down cages and recycling. Tesco moves thirty-two 45ft loads daily northbound on the Anglo-Scottish corridor, as well as a new service to Wales taking thirty-four 45ft boxes, while their service to the north of Scotland and the one from Tilbury take 22 containers each. Only Asda (not interviewed for this study) comes close, with

20 loads on the Anglo-Scottish route and 10 going to Aberdeen. At the time of writing, Tesco is planning to begin transporting up to 20 loads daily on the Aberdeen route, and more potential services are planned in collaboration with DRS, only one of which is likely to be a dedicated service. The additional Tesco volume means that the Aberdeen service is now fully utilised and will extend to seven-day operation. Indeed, DRS has noted that they have received additional interest from retailers due to the visible success of their Tesco trains.

The only other significant user of rail transport is wholesaler Costco, which sends 10–15 containers daily on the JG Russell service to Scotland. In the past, Costco put Aberdeen deliveries on this train (just to Coatbridge then by truck to Aberdeen), but, while the train arrives early enough to suit the central belt stores, there was insufficient time to drive from there to the Aberdeen store.

Only small numbers of containers are contributed by other users to the common-user Anglo-Scottish services. Sainsbury had used rail on some primary hauls to bring product from Scottish suppliers to their Midlands DCs, using the shared JG Russell service (although management of this flow has recently returned to the supplier). On the northbound route, Morrisons use the JG Russell service to move loads of picked pallets from Northampton to Bellshill. Morrisons had in the past trialled services between Trafford Park and Glasgow, and Coatbridge to Inverness. Both DHL for M&S and Waitrose use the WH Malcolm Anglo-Scottish service. M&S is building its own rail-connected DC at Castle Donington (see below for discussion). The Co-operative is currently running a trial on the WH Malcolm Anglo-Scottish service, taking two containers per night, five nights per week from the Midlands to their Scottish DC at Newhouse.[3]

Distribution Patterns

Primary distribution

Primary distribution refers to inbound flows into the DC. These flows can come from overseas through ports, the channel tunnel or by air, or they can come from within the UK. While the principal focus of intermodal transport for UK retailers is the movement of ambient grocery products on the domestic Anglo-Scottish route, consideration of imports through ports is required in order to understand how intermodal transport is based on both port and domestic flows. This combination makes matching of directional imbalances difficult and also raises complications with different wagon and container types, as discussed in the next section on operations. From a retail perspective, port flows are generally non-food lines such as clothes or electronics from the Far East moving through UK deepsea ports such as Felixstowe and Southampton. As the largest retailer, Tesco imports roughly 20,000 containers per year from the Far East through these two ports, the equivalent of about 400 loads per week.

---

3   Update July 2012. The retailer is now planning to double this trial to four containers per night.

Slow-moving food may be sourced from within the UK or from the continent. There are now significantly more European imports into Britain than in the past, with an increasing UK market for European wine and food. Fresh food is more time-dependent and is thus more likely to be sourced within the UK where possible, but with refrigeration technology it can be brought from the continent and even across the globe.

As discussed in the literature review, primary distribution has changed in recent times, with the introduction of DCs and PCCs further up the chain. This evolution has raised the possibility of the retailer extending control up the chain in order to manage flows more efficiently. Not all retailers have the resources or the desire to do this, as there are pros and cons to managing the primary distribution in-house or sub-contracting it to one or more firms. For instance, Sainsbury manages about 90 per cent of their inbound produce, 60 per cent of chilled and 10 per cent of ambient, whereas Tesco has more of a focus on primary distribution, with 60 per cent of ambient/grocery and 70 per cent of fresh flows moving through their primary network.

There is also the potential for an additional leg in the chain for the movement from a national distribution centre (NDC) to a regional distribution centre (RDC), depending on how the company has structured its distribution facilities. This will depend primarily on the product line, as there will tend to be more grocery/ambient DCs around the country to serve each particular area, whereas smaller flows such as frozen may only have one or two NDCs that must then filter product either to RDCs or direct to regional stores. For instance, Sainsbury has many composite DCs (chilled, ambient and fresh produce), whereas it has only two DCs for slow-moving food, health and beauty, and hazardous, two for frozen, two for clothes and only one non-food DC.

For Sainsbury, inbound flows are managed by the suppliers. They may occasionally haul a supply delivery, but generally only if they have a truck going somewhere and they need a backhaul. If Sainsbury only wants half a truckload, the supplier may prefer to take a whole truckload to a PCC, so Sainsbury will pick up that half load and leave the other half until needed. The inbound flows will be de-stuffed and stored at the DC then pallets will be picked from the racks to go to stores. Chilled and fruit will be cross-docked to go to the store the same day it arrives at the DC. A retailer will often only order small loads of exactly what they need. The transport cost may be higher, but they would rather pay more than have to store and manage extra product that is not yet required. Inventory management strategies (an entire subject that is beyond the scope of this research) thus exert considerable influence on transport decisions.

A wholesaler like Costco only stocks about 3,200 SKUs, so this is very different to a supermarket retailer, allowing Costco to maintain a far simpler operation. They have buyers overseas who make decisions for the whole global company (based in the US, Costco is one of the largest retail organisations in the world). The USA division will order so many containers of each product, as will the UK.

A unique aspect of primary distribution, in contrast to secondary, is the role of shipping lines. The UK is unusual in Europe in having a high proportion of carrier haulage (about 70 per cent), meaning that the shipping lines control distribution from the port to the inland destination. If moved by rail, it is usually with Freightliner on its ex port services, although DB Schenker has begun to compete successfully in this market. However, carrier haulage gives less control to the retailers. This is one area in which a large company like Tesco, with growing experience at managing its primary network, can negotiate port-only prices and manage the inland leg itself. The flows currently managed in this way are going by road, but plans are being developed to shift some of these flows (mostly Felixstowe/Southampton to Daventry) to rail.

Secondary distribution
Secondary distribution refers to the movement from the DC to the store. This move is more likely to be done in-house by the retailer or sub-contracted on a closer relationship. Tesco, Sainsbury and the Co-operative all run their own trucks for secondary distribution but will sub-contract occasionally where required (see Table 5.3).

Table 5.3   Distribution structure of each retailer interviewed

Company	Sector	Manage primary distribution	Manage secondary distribution	DCs
Tesco	Retailer	Partial – high	Yes	24
Sainsbury	Retailer	Partial – med	Yes, but about 50 per cent on third-party open-book basis	22
The Co-operative	Retailer	No	Yes	16
Costco	Wholesaler	No	Yes	1

The suitability of secondary flows for rail transport depends to a large degree on the distribution strategy of the retailer, such as which product lines are stored at the RDC and which require trunking from the NDC. When Tesco moves containers by rail from Daventry to Livingston, each container is designated for a specific store, with the relevant cages from Daventry inside. At Livingston additional cages are added to the container, which is then sent by truck to the store. This is done in the trunking station which is all cross-docked. Empty cages are returned to Daventry. Similarly, the Stobart Tesco train to Inverness takes boxes for specific stores which are then distributed by road by JG Russell, rather than being a DC to DC move.

An average Costco store has 4–5 full truck deliveries per day from the NDC and 20–30 small direct deliveries. As a wholesaler, Costco has a simpler operation for secondary flows. The majority of deliveries are overnight, to arrive at stores early in the morning (around 5am), which gives them time to get the pallets inside before store opening. Everything goes on pallets, which then go straight into the store. All their store deliveries are done by road except one rail service from Daventry to Coatbridge.

Tesco's large 1 million square foot DC at Livingston (opened 2007) is the only Tesco DC that has fresh, grocery, frozen, trunking and recycling all within the same facility. Around 4.5 million cases move weekly from here to about 250 stores across Scotland, north England and Northern Ireland. These 4.5 million cases cover grocery, non-food (picked at the DC in England then trunked up), fresh and frozen. Approximately one-third of product for NI comes through here: frozen and some picked slow-moving lines. There is a fresh (direct to this DC from suppliers) and a grocery depot in NI, but they are quite small.

There are 7,500 SKUs stored in the grocery part of the Livingston DC. The decision as to which lines should be picked at Daventry and trunked to Livingston and which should be stored there is monitored regularly, changing as different lines rise and fall in sales. All fresh food in Scotland moves through Livingston. The Livingston DC has around 900 trucks coming in and out on an average day, but this is an unusually large DC, as well as the fact that fewer of the trucks are double-decks, because many of the Scottish stores don't use them (partly because they do not have room to accommodate the larger vehicle, but also because they do not have enough demand to require it). Daventry has a higher proportion of double-deck trucks therefore fewer truck are required.

Lead time is an important aspect of planning all movements between DCs and stores. An ideal scenario, according to interviewees, would involve overnight picking and morning departure from the DC to reach the store by mid-afternoon, but the priority given to passenger trains during the day means that this cannot always be achieved. This means that intermodal growth is constrained by operating mostly at night, which requires stores to order from DCs in the morning so that the load can be picked in the afternoon, loaded at the DC at around 1600 to catch a 2000 departure on the train, which will then arrive at its destination in the early morning (0400–0500) for trucking to the store.

Centralisation and decentralisation

The literature review demonstrated that the leading retailers have pursued a process of rationalising the geographical coverage of their distribution facilities to improve the efficiency of their supply chains. Yet they cannot redesign their portfolio from scratch; legacy issues determine to a large extent where the DCs are located. Similarly, other models besides the generally-accepted centralised model may have some potential, such as port-centric logistics and continental hubs.

Asda and Tesco have both opened distribution centres at the port of Teesport in the northeast of the UK, focusing on imports. The interviews revealed that Tesco does not currently ship anything through this port, instead feedering containers from the ports of Felixstowe and Southampton. This result indicates that, even with port-centric strategies, centralisation tendencies are very hard to overcome. Tesco has fewer stores in the northeast than Asda so it might be concluded that the port-centric strategy was not suitable to their store coverage. This finding illustrates the difficulty in connecting port regionalisation strategies with inland freight systems.

Some interviewees questioned the operational aspects of port-centric logistics, suggesting that backhaul and container type issues may be difficult to overcome. When the DC is located in the port then imports arrive in maritime containers, which are emptied in the DC then the empty sent back to the shipping line. Picked loads are then distributed from the DC to the stores in 45ft trucks. The problem is that the only trucks coming to the port will be bringing maritime containers, so it can be difficult to match these flows, resulting in empty trucks coming to the DC to be loaded. Another disadvantage to the port-centric concept is that the company is anchored at that port with little option if a shipping line raises its prices or moves to another port.

Another distribution option in the UK with some potential is to make use of a continental hub to consolidate flows then bring them to the UK by rail or ferry. Tesco/Stobart work with 2XL in Zeebrugge to consolidate loads there (Red Bull, for example). French wine used to come in full loads; now Tesco de-stuffs them at Zeebrugge and consolidates many different loads into one container which can then go direct to the store in the UK, reducing their stock holdings from six weeks to one. However, there are operational reasons why this model has only niche attraction. The rates charged to transit the Channel Tunnel are considered by some interviewees to be high, part load patterns are complex, and the ferry also has constraints such as time, frequency and imbalance of flows.

Tesco is also exploring the possibility of consolidation centre use in China. Instead of bringing full containers of each product to the UK which then must go to the NDC for de-stuffing, they could put a mixed load in the container before it leaves China. It might then be able to go straight to a regional DC or even a store, thus removing a leg from the UK distribution. This could have more potential to change distribution patterns. If all importers of freight from the Far East picked loads there before sending them to the UK, the centralised pattern of DCs in the UK could be altered. However, as these products are mostly non-time sensitive products, they do not constitute the majority of product lines discussed in this chapter.

Operational Issues

Rail operations
Loading gauge restriction on the UK network is a well-known issue, mostly in the north of Scotland and on the East Coast Mainline (ECML), which is used when services are diverted from the UK's primary rail freight corridor, the West Coast Mainline (WCML). This is generally more of an issue for maritime containers coming through ports, as these are gradually moving towards a majority of high cube (i.e. 9ft6 height rather than the old standard of 8ft6). While the major parts of the network (Felixstowe and Southampton to the Midlands and thence to

Scotland on the WCML) can now take these containers on standard wagons (W10 loading gauge), significant portions can only take them on special low wagons (W9 or W8 gauge). Containers designed for purely domestic flows (e.g. the Tesco rail containers designed in conjunction with Stobart Rail) are more likely to be standard 8ft6 height, thus avoiding this problem.

These constraints represent a specific challenge for intermodal transport of deep sea containers, as many routes linking ports to inland terminals have not yet been upgraded to W10 gauge. Moreover, high cube containers are expected to increase to 65–70 per cent of the market by 2023 (Network Rail, 2007; Woodburn, 2008). As the cost is too prohibitive to raise the required bridges to allow high cubes through on some routes, the only feasible option is to use low wagons; a representative comment was that 'a wagon solution would be more cost-effective'. However, purchase and maintenance of specialist wagons is typically more expensive and they reduce capacity, with a maximum capacity of a single 45ft container per 54ft wagon instead of a standard 60ft wagon used for port flows, which can take combinations of 3x 20ft or 40ft + 20ft containers. This makes low wagons economically undesirable for freight operators (Woodburn, 2008; Network Rail, 2007). There is also the problem of a mismatch of wagon configurations for operators wanting to run both port and domestic routes.

Lack of visibility and knowledge have both been cited as key barriers to modal shift, in conjunction with lower flexibility, frequency and reliability than road. While flexibility and frequency are difficult to alter, the perception in the industry regarding reliability is changing. While some reports still claim a negative perception among small hauliers (RHA, 2007), all rail users interviewed for this research stated that their use of rail had been reliable. In particular, rail had proved more reliable than road during the hard winter in 2010/11, when some roads were closed due to snow and ice. The feeling is that, as shippers gain experience using rail, they know that they can contact a freight forwarder or rail operator and put even a single container on a timetabled rail service. However, work is required to achieve this position, and to extend it will require further work on behalf of 3PLs who can provide a door-to-door solution to customers, aiming to emulate the responsiveness of a road haulier.

It has been suggested that the UK suffers from an insufficiency of intermodal terminals, in particular that many smaller sites and sheds should be rail connected (SRA, 2004; Berkeley, 2010). Access is also an issue; the Office of Rail Regulation (ORR) is pursuing an investigation into whether withholding of terminal access by incumbent operators is anti-competitive (ORR, 2011). 3PLs interviewed suggested that a strategy of more rail-connected warehouses is less desirable and the major intermodal terminals are sufficient in the short term.

According to interviewees, asset utilisation is more important than break-even distance, even if comprised of a number of short-distance services. 'Beware of management accountants', cautioned one interviewee, because they look at the individual costs of running a train, but they do not factor in operational aspects like utilisation and cross subsidy across their service portfolio. Daytime running is generally possible in Scotland because the lines are not as congested, but in the rest of the UK most freight trains run at night due to path restrictions during the day. This means that they often sit idle all day. The approach taken by rail operators is that if a train can be kept running most of the day then it will make money, so if a train is standing idle then any service, no matter how short, is worth running. This marginal cost approach explains the short service between Elderslie and Grangemouth run by DRS.

Handling charges involved in changing between modes have always been a barrier to greater use of intermodal transport. However, it was found in the interviews that the price paid by shippers for handling is something of a contentious topic. Users feel that they are being given a nominal price to pay, without evidence of a relation to the actual cost to the terminal operator of providing this service. Tesco has been able to bargain this price down, but rail operators feel that they cannot go any further or they will not be able to provide the service. Moreover, this is related to a wider issue of the visibility of the true cost of rail movements. One interviewee suggested that the quote they are given is simply based on being 'slightly cheaper than road' rather than being directly related to the actual costs of the rail service. Some shippers say that they would like greater visibility of the cost to the provider of the entire rail service, including the trunk haul, so that they know the basis for the price. This is similar to the greater control over primary distribution sought through the use of factory gate pricing. They are both strategies for removing the need for the retailer to pay a profit margin on top of the base cost of the transport service.

The difficulty of obtaining backhauls is another significant aspect of the economics of rail. Rail is generally considered to be competitive with road if there are flows in both directions, therefore the backhaul is often the key issue in making intermodal transport work. By integrating its primary and secondary distribution, Tesco has been able to match supplier deliveries inbound to the Daventry NDC with outbound distribution to RDCs. Tesco sells space on its dedicated trains to its suppliers, thus inserting itself in a chain of vertical cooperation that draws the rail operator, 3PL and supplier together.

More backhauls from Scotland to England are needed to support the Anglo-Scottish services. Chilled trucks returning from Scotland are currently backloaded with fish and meat. One interviewee noted that it can be difficult to get ambient backloads from Scotland, while another felt certain that the loads are there, saying that 'it is just a matter of making it work', sometimes just persuading a company that has not used rail before to run a trial. It was suggested that there are many companies with a few containers a day that could use it, or that may require consolidation of LCL before sending them south by rail. These interview findings suggest that consolidation could be a key issue to promote further use of intermodal transport and to integrate road and rail more seamlessly, and could be pursued in future research.

As well as rail operations, road haulage must also be understood in order to contribute to supporting the growth of intermodal transport. Road is built into supply chains because of its inherent qualities. For example, one retailer pointed out that 'staggered delivery is easier to manage'. For example, if 30 containers arrive together on a train then it can potentially cause congestion at the DC.

Wagons, containers and retail cages
The key issues in terms of container dimensions relate to pallets and retail cages. A standard truck will take 26 UK pallets (or 52 if pallets are loaded less high so they can be stacked double). The internal width of a truck is 2.48m, which means UK pallets can fit 2× 1.2m across, whereas an ISO container has an internal width of 2.35m which prevents this configuration. Whether a shipper is using UK or Euro pallets is a big issue in terms of getting the maximum value from a container. UK pallets (GKN or CHEP) measure 1200mm × 1000mm, thus fitting into a trailer more efficiently, whereas European pallets measure 800mm × 1000m. A road trailer takes 26 UK pallets or 33 Euro pallets, whereas a 40ft maritime container takes only 22 UK pallets or 25 euro pallets.

Imports from the Far East are generally not palletised; the contents are loaded loose or break bulk in the deepsea container. This will be de-stuffed in the importing country and then put into a trailer for domestic movement. For intra-European loads, this is not practical therefore 'pallet-wide' containers have been developed on European short sea routes. These are 2.4 inches wider than standard containers (i.e. just over 8ft2 rather than 8ft wide), giving an internal width of 2.44m, similar to a road trailer. A pallet-wide 40ft container will take 30 euro pallets rather than 25 in standard width or 33 in a road trailer. 45ft length is the preferred option, taking the same pallet loading (UK or Euro) as a road trailer.

There is a move in Europe to make 45ft pallet-wide maritime containers the industry standard (Bouley, 2012). Problems exist because most deepsea ships cannot accommodate these containers in their cellular holds and EU directive 96/53/EC forbids standard 45ft long containers on trucks (although modified designs with chamfered corners can fit within this limit).

The next issue is retail cages. Retail movements to stores are generally done in cages, and greater economies can be achieved by transporting these in double-deck trucks, which are almost unique to the UK (McKinnon, 2010). A standard truck takes 45 retail cages, as does a 45ft rail container, whereas a double-deck truck can take 72 cages. One retailer commented that, 'because we run double-deck road trailers, it is difficult for the rail operators to compete on price'. Double-decks currently form about 20 per cent of the Tesco fleet, and may eventually rise to around half.

Domestic intermodal containers used in the UK by 3PLs such as Stobart, WH Malcolm and JG Russell are 45ft pallet-wide to deal with the issues noted above. The Stobart/Tesco containers are curtain siders, which is common on trucks but not on trains. However, they can be targets for thieves, as trains are often required to stop on the line during the night. Additionally, curtain siders cannot be stacked, like swap bodies but unlike ISO maritime boxes. Moreover, HGVs can be compartmentalised for chilled, frozen and ambient product but current rail

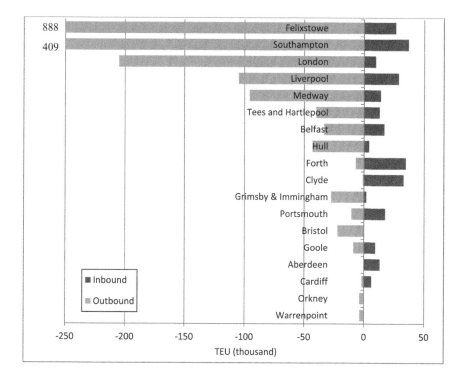

Figure 5.2   Empty container movements through UK ports, 2010
Note: Felixstowe and Southampton have been truncated
Source: Author, based on DfT (2011)

containers cannot be, limiting their flexibility. All of these operational issues can be potential disincentives to modal shift to rail.

Transporting high-cube containers (9ft6 high) on non-gauge-cleared routes means using special wagons such as Megafrets and Lowliners, which are more expensive to buy and to maintain. In addition, these wagons are generally 54ft and carry 45ft domestic boxes, meaning 9ft of length is wasted per wagon. This is now being addressed by new low 45ft wagons. There is also the conflict between port flows (generally hauled by Freightliner on 60ft wagons to cater for 20ft and 40ft containers) vs domestic flows (which are primarily 45ft boxes).

It was noted in the discussion of primary and secondary distribution that, for a large retailer like Tesco, managing both legs allows the matching of flows to increase the economic viability of a service. This strategy is challenged by the acute container imbalance on the Anglo-Scottish route. Figure 5.2 shows that, while the large deepsea ports export empty containers back to the Far East, smaller ports, particularly in the north, are required to import empty containers to fill with exports.

Northbound imports to Scotland come mostly as 45ft pallet-wide road trailers or swap bodies (and now rail containers), as they are retail and other movements from DCs in the Midlands. The majority of Scotland's exports leave as 20ft/40ft maritime containers, either through ports or on rail. This container mismatch also affects wagon configurations, for example sometimes 45ft containers are carried on 60ft wagons designed for 20ft/40ft combinations.

It was found in the interviews that industry discussions have taken place with regard to the possibility of a joint action between retailers (northbound 45ft boxes) and whisky producers (southbound 20ft/40ft boxes). If both were to use the same type of container and transload the contents at one end, the problem could be resolved, as long as the savings made from matching the flows outweigh the cost of transloading. However, distillers understandably do not want their high value cargo to be handled any more than is necessary, and retailers are lacking an incentive to inconvenience their operations in order to reduce the repositioning costs paid by Scottish shippers.

Integration and Collaboration

Vertical

The most conspicuous vertical collaboration observed in this research was between retailer Tesco, logistics provider Stobart and rail operator DRS. A close working relationship has enabled all parties to develop a knowledge of requirements and adjust operations to support the collaborative endeavour. They work together to develop new services, discuss requirements and solve operational issues. A special container has been designed to transport Tesco flows by rail. While the original blue containers were branded 'Tesco Less CO_2', the new generation of containers do not display the Tesco brand, as part of a move towards the possibility of transporting loads from other shippers in these containers. However, it is unclear whether competitors would use these containers which retain their brand identity in the well-known blue boxes.

Vertical collaboration or integration between all levels of rail operations (from terminal operation, traction provision, train management and road haulage) can be pursued in a variety of ways (this point will be expanded in the discussion in Chapter 7). Terminals can be run by rail operators (e.g. Freightliner or DB Schenker), 3PLs (e.g. WH Malcolm) or other companies (e.g. ABP at Hams Hall), or even be private sidings for which the operation is sub-contracted (e.g. Stobart operating the Tesco shed at DIRFT). Likewise, the customer side of trains is normally managed by a 3PL rather than the rail operator (e.g. JG Russell, WH Malcolm and Stobart operating trains with traction provided by DRS or DB Schenker), but for other trains the management is also done by the rail operator (e.g. Freightliner or DB Schenker). Varying levels of vertical cooperation and integration can therefore be identified, depending on the particular service which may or may not be running into the terminal of the rail operator.

Whether a 3PL is going to operate another terminal depends largely on the structure of their existing business. It would need to fit in with their other operations; for example, the location of their existing haulage and distribution customers. Whether they would move into direct rail operation is a more complicated question as a whole new level of expertise is required and many barriers make market entry difficult (Slack and Vogt, 2007). However, it is interesting that the three 3PLs involved in intermodal transport are smaller than the huge global businesses that are predominantly road-based, such as Wincanton and Norbert Dentressangle. These larger companies tend to focus on running distribution centres and supply chain logistics rather than intermodal transport.

Horizontal
Turning to horizontal collaboration, this can relate to retailers sharing space, either within containers (or using each other's containers), or sharing space on trains. The former does not currently happen, but the latter is already in evidence, with most retailers using common-user 3PL trains as noted earlier. Interviewees have recognised that this may be the only way to make services viable on some routes. It can also refer to 3PLs sharing space on their services, which does already happen. It is sometimes on an ad hoc basis, but also on a regular basis, especially boxes coming from ports, as these services are mostly run by Freightliner who specialises in these flows. 3PLs have not entered this market so they will buy space on those trains (e.g. Stobart bringing boxes from Tilbury to their hub at Widnes). 3PLs can collaborate in other ways; for example, Stobart runs the Tesco train from Mossend to Inverness, where it terminates at the JG Russell terminal, from which point JG Russell distributes the containers to stores by road. Likewise, the Stobart Valencia fruit train utilises the JG Russell terminal at Barking.

Tesco is currently the only retailer large enough to fill a complete train. When planning a new service, they must decide if they are going to operate a dedicated service, in which case the retailer commits to paying for the whole train. In this case, they must take responsibility for filling any empty wagons or risk losing money. Scheduled services are less risky as they may be used by any shippers (Lammgård, 2012), but having a dedicated train is attractive because it grants more control over the timings and operation of the service. Such a scenario enables the retailer to plan the primary and secondary distribution as part of a unified system. Rail can only work if both primary and secondary distribution works together (i.e. backhauls), but this needs to be managed.

In an ideal scenario, all retailers would prefer to have their own rail-connected warehouses rather than using a shared terminal to load a common-user train. This means that they would need to run a full train, which would reduce opportunities for collaborating and perhaps shows that they are not so keen on sharing space. Retailer M&S is building a new rail-connected DC at Castle Donington, but unless the retailer can provide enough volume for regular services (which is unlikely), servicing this terminal will make asset utilisation more difficult for rail operators.

In order to work successfully an operator (or someone else) will need to provide more rail flows to this terminal to get better asset utilisation from the rolling stock.

The primary rail hub in the UK is Daventry, but within the site there is the option of using the common-user terminal operated by WH Malcolm, or private sidings if a company has selected one of the sites with private sidings. This is still a feasible option, but it is representative of the desire for supply chain control evidenced by the large retailers who would prefer not to share third-party trains; as shown above, other than Tesco, all retail use of trains is on third-party services.

Other interesting collaborations occur as responses to unusual circumstances. In the winter of 2010/11 the UK experienced severe disruption due to heavy snow; 3PLs needed to move cargo so they put together a train with DRS traction, Freightliner wagons and WH Malcolm traffic. The difference is that while 3PLs will share space when needed, they do not actually run a regular train together. It could be that in the future this kind of collaboration will be required to improve the economics of rail operation.

Role of Government

As noted earlier, Network Rail, a nominally private but government-owned company, owns and operates the track infrastructure, with intermodal terminals owned or leased by private operators. Private rail operators compete to run services. Therefore the role of government in intermodal transport development in the UK is limited.

Network Rail has identified a Strategic Freight Network (SFN) (Network Rail, 2008; DfT, 2009). Investment is organised in 'control periods', and the UK government committed £200m towards the SFN in the current Control Period 4 (2009–2014). Specific goals include longer and heavier trains, efficient operating characteristics, 7-day/24-hour capability, W12 loading gauge on all strategic container routes, European (UIC GB+) loading gauge from High Speed 1 (HS1) to the Midlands, increased freight capacity, electrification of freight routes, development of strategic rail freight interchanges and terminals and protection of strategic freight capacity. Capacity issues are not generally a problem in Scotland or from Scotland down to Crewe. Capacity problems exist from the Midlands down, primarily during the day.

The recent upgrade of Southampton rail line to take high cubes has already yielded significant results, as DB Schenker secured a deal with CMA CGM to take 25,000 boxes by rail to inland terminals at Birmingham, Manchester Trafford Park and Wakefield (Hailey, 2011). Similarly, Network Rail has recently upgraded to W10 the route from Felixstowe to Nuneaton, meaning W10 trains to the north no longer need to divert through London. The route through the Channel Tunnel and up to Barking is cleared to European gauge.

Network Rail, according to the interviewees, is much better to work with than previous infrastructure owner Railtrack, but more flexibility is required in some areas. For example, incumbent operators are loath to give up paths even if they do not use them, and they are only required to run a train once a year to retain the path. It was also suggested that some paths are, in reality, a higher loading gauge than listed, and this could be cleared up with 'only some paperwork' and a trial run. Another interviewee complained about having to pay double for a terminal to open on a Sunday; increased Sunday operations would be welcomed by shippers, particularly retailers, as indeed would permission for night time deliveries to stores.

Government grants (FFG for infrastructure and MSRS for operating subsidies) have been used to support the shift of retail (and other) flows from road to rail. While FFG infrastructure grants were introduced in the UK in 1974, rail operating grants began with the TAG scheme set up in 1993 by the UK DfT, with the Scottish portion transferred to Scotland in 2001. TAG was complemented by CNRS from 2004 to 2007. Both of these were replaced by REPS (2007–2010) and then MSRS (since 2010).

Woodburn (2007) reported on previous assessments of the FFG system in the 1990s (NAO, 1996; PAC, 1997; Haywood, 1999) as well as assessing the grants from 1997 to 2005. He made an interesting comparison with continental European countries, where national rail operators are still subsidised at a general level, suggesting that the liberalisation of this market will invalidate this approach and require a similar system to the flow-based FFG awards in the UK. He concluded that the FFG system had been largely successful, providing on average two-thirds of the required funding for the facilities, meaning that one-third of the cost was paid by the private operators in situations where these facilities may not otherwise have been built. He found that the vast majority of these grants can be considered successful, with the failed projects attributable mainly to 'company or supply chain changes that were not foreseen at the time of the award' (p.325).

Table 5.4 (below) lists FFG funding for intermodal terminals in England and Scotland. This does not include other rail investments such as bulk terminals, rail-connected warehouses or wagons. For example, Stobart received £200,000 in 2006 from the Scottish Government towards the Tesco containers and a further £525,000 in 2008 towards the Grangemouth to Inverness Tesco route.

The table shows that many of the intermodal terminals discussed in this chapter have received infrastructure funding through the FFG programme. The funding has supported intermodal development in others ways, such as subsidising the construction of the Tesco/Stobart rail containers. More money has been spent in Scotland on intermodal terminals, while many grants in England and Wales have gone to other operational requirements, as well as some large intermodal projects in ports.

Interviewees were all supportive of the government grants and critical of their reduction (Scotland) and/or removal (England and Wales). Having said that, some concerns were raised that the FFG system could have been used more strategically and that the process deterred some projects that might have been successful. It must, however, be recognised that the economic and operational realities of the freight business can make it difficult to use such funding strategically (e.g. by using the planning system to designate strategic terminals via a top-down process

Table 5.4   List of FFG awards in England and Scotland to intermodal terminals

Company	Terminal	Year	DfT	Scottish Government
Freightliner	Trafford Park	1997	£723,00	
JG Russell	Deanside	1997		£3,045,000
TDG Nexus (site now owned by DB Schenker)	Grangemouth	1999		£3,233,000
Potter Group	Ely	2000	£373,046	
WH Malcolm	Grangemouth	2000		£246,000
Potter Group	Ely	2001	£101,000	
Freightliner	Leeds	2002	£196,656	
Potter Group	Selby	2002	£1,579,051	
WH Malcolm	Grangemouth	2002	£582,602	
Roadway Container Logistics	Manchester	2002	£328,350	
ABP	Hams Hall	2002	£1,192,965	
PD Stirling	Mossend	2002		£1,878,300
EWS (now DB Schenker)	Mossend	2003		£654,000
WH Malcolm	Grangemouth	2003		£882,000
ARR Craib	Aberdeen	2004		£144,546
WH Malcolm	Grangemouth	2004		£137,678
WH Malcolm	Elderslie	2005		£1,647,000
WH Malcolm	Elderslie	2006		£572,000
JG Russell	Coatbridge	2008		£1,842,617
Total			**£5,076,670**	**£14,282,141**

Source: Author, compiled from data from DfT and Scottish Government

rather than relying on ad hoc funding applications). When questioned about other government incentives for modal shift, while interviewees considered it unfeasible to enforce use of rail, the possibility that the Department for Transport (DfT) could in future allow overweight trucks between DCs and intermodal terminals was felt to be more realistic.

Table 5.5 lists the recipients of MSRS operating subsidies in 2010/11. The table reveals that the key players discussed in this chapter have received significant operational funding for their intermodal services. Unless the economic competitiveness of rail is improved, these operational subsidies are likely to be

Table 5.5   Recipients of operating subsidies through MSRS intermodal 2010/11

Recipient	Transport Scotland	DfT	Total
DB Schenker	£33,994	£192,749	£226,743
Direct Rail Services	£310,676	£678,817	£989,493
Eddie Stobart	£308,113	£328,209	£636,322
Freightliner	£27,977	£56,190	£84,167
JG Russell	£136,157	£752,158	£888,315
Total	£816,917	£2,008,123	£2,825,040

Source: Author, compiled from government sources

ongoing, and if they are removed then some of the services discussed in this chapter could cease to operate. MSRS is an ongoing subsidy, whereas the water-based equivalent subsidy (WFG – waterborne freight grant) is only available if the service can be shown to become feasible within three years, hence the lack of success with that funding stream. FFGs need to be tied to a particular flow, therefore some of the above grants may be for non-intermodal traffic but be useful for future traffic flows. It is difficult to be strategic when there are issues such as lead times, purchase or lease of locomotives, wagons and so on.

Another issue that has not yet been covered in this analysis is the misalignment of approach between track and terminals. Network Rail has identified a 'strategic freight network' of core and diversionary routes in England and Wales on which to focus investment. Transport Scotland has not yet created such an official route in Scotland, but it is being proposed for Control Period 5 (2014–2019) (Network Rail, 2011). So there is a strategic plan for the infrastructure network (as this is managed by Network Rail) but not for terminals (as these are privately operated). These are developed through ad hoc funding applications without strategic focus. Moreover, an operator has to identify a road flow, set up a service to shift it, then apply for the funding and then develop the terminal facilities, by which time the flow may have evaporated.

It could be more appropriate to look at this problem strategically. One option could be to merge the FFG funding into the strategic Network Rail programme, thus considering terminals as infrastructure, as some authors have suggested (Woxenius and Bärthel, 2008). Some of the budget for the Strategic Freight Network could be made available for terminal upgrades. Network Rail could put out a call for terminal operators to apply, based on their previous flows and business, current usage, upcoming business, etc. This could be supported like a TIGER grant (see discussion of funding in the USA in Chapter 6) by a public body and even combined in a package with relevant network upgrades into a corridor approach, which would be a suitable role for a regional body such as the RTPs in

Scotland, which are currently under-used. This is easier in the United States with vertical integration linking track, terminal and operator, but even if that system cannot be replicated, a strategic system based on invited applications might work better than the current ad hoc system.

While issues have been noted above with the current grant system, recognised issues with attempting to make intermodal traffic economically competitive with road must be accepted and will not go away. Otherwise it is more of a structural issue of consolidating traffic at large sites to achieve economies of scale, removing the last mile where possible, and other operational issues such as asset utilisation, as discussed above. Large shippers could be incentivised to locate at these sites, otherwise ongoing operating subsidies will be required in perpetuity. Also, when it comes to a discussion of infrastructure subsidies versus operating subsidies, the European case studies and interviews have revealed the difficulties of this problem. Most grants are for infrastructure and terminal development, yet due to the economics of transport services they cannot compete with road and therefore struggle to get traffic.

Analysing the Case Study

Table 5.6 (below) sets out the thematic matrix with the relevant data noted against each factor. The table is followed by a discussion of these findings.

While the majority of the findings from the British case corroborate the previous research discussed in the literature review, some new findings can be generalised to comparable market sectors within the UK as well as other regions with similar market and geographical characteristics, such as continental Europe.

The centralisation of DCs identified in the literature was confirmed by the analysis of the spatial development of the retail sector in this chapter. Some potential drivers for decentralisation were identified, such as port-centric logistics and continental hubs. These may be classified as regionalisation strategies through which ports attempt to control hinterland links by embedding themselves in distribution chains, but the analysis in this chapter has suggested that they have only limited potential. This observed centralisation is a key component of successful intermodal logistics, as it has facilitated trunk hauls between NDCs and DCs; however, different practices have been observed in the present research. For example, sending a partially-full container from the NDC to the Scottish DC where further cages are added in the trunking station, and thence to the store (e.g. Tesco) is different to sending a full container from Midlands NDC to Scottish rail terminal and direct to store (e.g. the Co-operative).

The rail sector analysis revealed that the intermodal market is growing, but while healthy competition is observed between three 3PLs, traction for all but one of the services is provided by one operator, thus suggesting that it is not an easy market to enter. Only one retailer (Tesco) has significant volume, although Asda is making efforts with both rail services and their port-centric import centre, due partly to their greater concentration of stores in the northeast. Interviews revealed that Tesco does not actually ship anything through Teesport at the moment which indicates that, even with port-centric strategies, centralisation tendencies are very hard to overcome. Thus port regionalisation is very difficult to connect with inland freight systems.

The key route is Anglo-Scottish, though port flows are relevant, for example Tesco seeking to replace carrier haulage with their own primary network. Intermodal terminals for these flows at the corridor termini at England and Scotland were identified along with current service provision. The analysis also revealed that government funding has been essential in upgrading freight handling facilities at many UK intermodal terminals; moreover, annual operational subsidies of £2.8m underpin the services discussed in this chapter.

Turning to the operational perspective, the case study showed that asset utilisation is key for rail operators as expensive assets are forced to remain idle while daytime paths are used by passenger trains. Other known operational issues play crucial roles, such as wagon and container management; it was seen in this case how the container imbalance on the Anglo-Scottish corridor increases the difficulty of sourcing backhauls which are essential to the economic viability of these services. This imbalance in both flows and equipment is exacerbated by the lack of coordination between domestic and port movements. These flows are generally managed by different parts of the retail organisation (e.g. grocery focus on domestic flows and general merchandise for port flows), but intermodal transport efficiency requires that they be considered together, and this is required in turn for successful port regionalisation to occur.

Successful intermodal transport requires full trains; horizontal collaboration can help achieve this goal, but while such collaboration is occurring to some degree between 3PLs, it is not happening with retailers. The case analysis revealed that retailers prefer private sidings rather than common-user terminals; this preference splits economies of scale and can be a barrier to greater use of intermodal transport. Public planners should, therefore, consider whether common-user platforms should be preferred in the planning system rather than more rail-connected private sites. Vertical integration is more common than horizontal, as was to be expected. The literature hinted at increasing collaboration but did not elaborate, therefore the specificity of this case provided an opportunity to understand this process in greater detail. There was no evidence in the case of a change to the current lack of horizontal collaboration, which confirmed the literature. Increasing vertical collaboration may be possible, as other retailers see that this is the only way to make intermodal transport work. But actual integration or expansion beyond an organisation's core business is not in evidence.

The research approach pursued in this chapter has demonstrated the value of taking a broader approach to the support of intermodal service development, by studying the role of 3PLs as well as retailers. In the interviews, 3PLs raised the importance of consolidation centres for converting LCL into FCL which can feed intermodal services if they are located at intermodal hubs, in particular southbound flows on the Anglo-Scottish route to provide backhauls for northbound retail flows.

Table 5.6 Applying the thematic matrix

Factor	No.	Sub-factor	Data
	1	Primary distribution	See Table 5.3 for detail on each retailer. In most cases primary distribution is managed by the suppliers, although this involves collaboration with retailers for delivery times, sizes, etc. FGP was not considered attractive in most cases, with the exception of Tesco, most likely being the only retailer large enough to practice FGP.
	2	Secondary distribution	See Table 5.3 for detail on each retailer. All three retailers and the wholesaler manage their secondary distribution and in conjunction with the discussion in the literature, this is considered the dominant paradigm.
	3	Distribution centres	The literature and the interviews confirmed that retailers have been rationalising their distribution centre locations and types, reducing numbers and moving more towards composites. This process continues. These DCs are larger than before and purpose built to fit in with detailed planning of flows and different product types and combinations. Very few have rail connections.
	4	Centralisation or not	Centralisation remains the dominant paradigm, and little interest was evinced in the interviews or in observed practice to decentralise this system. Some uses of port-centric logistics and continental hubs were revealed, but the latter is only for some niche products like wine, whereas the former has operational limitations as discussed in this chapter. Asda's PCL operation at Teesport may be successful but without interviewing them it is difficult to discuss that further. Tesco's similar operation seems of doubtful success as it doesn't suit their store profile and due to their deals with shipping lines they no longer import their containers through Teesport anyway. In terms of decentralising DCs to the north, the limit has already been reached, with a major DC for each retailer in Scotland already. Flows to Scotland are now composed of stock from Midlands DCs and some SKUs from Scottish DCs, but fully independent Scottish DCs are not going to be feasible.
	5	Product and route characteristics	The primary domestic corridor is the Anglo-Scottish route, where domestic ambient product moves northbound in 45ft pallet-wide containers, backfilled with a variety of loads, from suppliers and other consolidated loads. The port flows are palletised or break bulk general merchandise in 20/40ft deepsea containers.
	6	Intermodal transport	Only one large retailer (Tesco) is making significant use of intermodal transport, partly because it is large enough to fill trains, but also because of a decision to pursue it, although it was unclear to what degree this is a company decision or a personal decision on the part of the staff. All other users are only dabbling, although wholesaler Costco has a regular flow of reasonable significance. On the whole, however, transport is secondary to logistics, and is thus based on road transport. In the interviews, 3PLs seemed confident that intermodal use will increase as it becomes more normal and well understood (in addition to other drivers like fuel price), but while retailers say they are positive about intermodal, actions show that it remains a minority interest, and operational issues (discussed below) suggest that it will remain so at least in the short to medium term.

Spatial distribution of the market

7	Logistics	Logistics decisions take precedence over transport. DCs are designed to support logistics and supply chain requirements, located for road access rather than rail, and transport must fit in with product management. There are few signs of logistics being integrated with transport. Even Tesco, the leader in intermodal use, built their Scottish DC at Livingston right next to a rail line but did not build a rail connection.
8	Other	
9	Infrastructure	Rail infrastructure is owned by government (indirectly through Network Rail). Periodic investment takes place (e.g. Strategic Freight Network). Upgrades are needed to increase train length as well as to increase capacity through more passing loops and double tracking. The high-cube issue remains on some routes, but main routes are increasingly cleared to W10. Capacity is generally reasonable in Scotland and north England (although high-cube issues remain on lines to the north of Scotland). Capacity issues during the day in south England limit paths for freight trains. Government investment is ongoing but daytime capacity issues will not be resolved.
10	Operations	Difficulties were raised in interviews regarding backhauls (which are essential for making routes viable) and the need to cross-subsidise services through high asset utilisation (which is difficult when daytime paths in England are not available). Setting up a new service is costly and has significant lead time. The image of rail among users has improved and all shippers in this research said they were happy with the reliability of rail. However, rail is inherently limited in its flexibility.
11	Operators	See Table 5.1 and Table 5.2. All but one service on the Anglo-Scottish route used by retailers is provided by one rail operator, although other intermodal services exist in England. There remains a separation between port flows (Freightliner and DB Schenker) and domestic (DRS). Carrier haulage is unusually high in the UK compared to the Continent, which may explain this disjunction.
12	Equipment	Numerous equipment issues constraining greater development of intermodal transport were identified in interviews and the document analysis. Northbound retail movements of 45ft pallet-wide containers do not balance with southbound whisky movements in 20/40ft deepsea containers. 45ft pallet-wide European containers could help this problem, but no evidence has been found of movement on this issue. The high-cube clearance issue was already raised in point 9. Lack of clearance necessitates low wagons, which are generally 54ft, thus wasting space, and these wagons do not match with 60ft wagons used for port flows, further embedding the disjunction between port and domestic flows. Wagon and container management play crucial roles, and the container imbalance on the Anglo-Scottish corridor increases the difficulty of sourcing backhauls which are essential to the economic viability of these services.

Operational rail issues

(continued ...)

Factor	No.	Sub-factor	Data
Integration and collaboration	13	Price	Interviews revealed that the price paid by shippers for handling is a contentious topic. Users feel that they are being given a nominal price to pay without evidence of a relation to the actual cost to the terminal operator of providing this service. Tesco has been able to bargain this price down, but rail operators feel that they cannot go any further or they will not be able to provide the service. It was suggested by one retailer that the quote they are given is simply based on being 'slightly cheaper than road' rather than being based on the actual costs of the rail service. Some shippers say that they would like greater visibility of the cost to the provider of the entire rail service, including the trunk haul, so that they know what the prices are based on. Published tariffs would simplify the process but this is unattractive to operators who would lose influence this way, similar to suppliers losing the ability to cross-subsidise their profits through transport costs when retailers force them into FGP contracts.
	14	Other	
	15	Horizontal	Very low among retailers and little appetite to change this, which could be a significant barrier to greater use of intermodal transport. 3PLs do collaborate with each other sometimes. Rail operators do so only occasionally to solve operational issues but it is not a regular action.
	16	Vertical	Vertical collaboration is more common than horizontal, as it is necessary in the modern complicated logistics and transport environment. In particular for use of intermodal rather than road transport, high vertical collaboration is essential. Therefore it is Tesco/Stobart/DRS who are collaborating the most because they are the main intermodal configuration. But only collaboration rather than actual integration.
	17	Consolidation	Little appetite for third-party consolidation, which relates to the lack of horizontal collaboration observed. 3PLs were more alert to this requirement than retailers, which perhaps relates to the higher incidence of horizontal collaboration among the 3PLs. In general, the 3PLs seem more focused on solving operational issues while retailers are more concerned with managing their own business rather than altering it to suit larger collaborative interests such as intermodal transport requires.
	18	Evidence of change	There was no evidence in the interviews of a change to the current lack of horizontal collaboration. However increasing vertical collaboration may be possible, as other retailers see that this is the only way to make intermodal transport work. But actual integration or expansion beyond an organisation's core business is not in evidence. For example, none of the 3PLs were interested in becoming a rail operator, although some of them handle trains at their terminals. Any future change would appear to relate purely to a deepening of existing vertical collaborations.
	19	Other	

Role of government in intermodal transport	20	Policy	Some ideas, such as government legislation for taxing C02, enforcing x per cent use of rail or x per cent maximum empty running, were discussed by interviewees but considered unrealistic. It was considered more feasible that the DfT could in future allow overweight trucks between DCs and intermodal terminals as a way of stimulating intermodal transport. Generally, however, the role of government was considered more in terms of infrastructure upgrades than direct intervention in operations.
	21	Planning	The UK planning system is important for developing new or expanding current intermodal terminals. This was not a major aspect of the interviews so it was only touched on. It was discussed by some interviewees in the context of third-party or single-user sites. The DIRFT terminal is currently common-user, although some shippers located there, such as Tesco, have their own rail connections. Concern was expressed by some interviewees that planning consent for more single-user sites could split scale economies and make rail more difficult, although the problem is that shippers prefer their own connections rather than sharing either trains or terminals. In terms of planning specifically for the rail network, this is done by Network Rail through their 'control periods', each of which schedules upgrades on identified sections of track, but these are subject to grants from the national government.
	22	Funding	Data showed that government grants (FFG for infrastructure and MSRS for operating subsidies) have been instrumental in supporting the shift of retail (and other) flows from road to rail. Many intermodal terminals in the UK have benefited from FFG funding at one time or another, most actors in this chapter receive ongoing operational subsidies and the funding has supported intermodal development in other ways, such as subsidising the construction of the Tesco/Stobart rail containers. Interviewees were all supportive of the government grants and critical of their reduction and/or removal, although there were some concerns that the FFG system could have been used more strategically and that the process deterred some projects that might have been successful. Generally, however, the role of government was considered more in terms of infrastructure upgrades than direct intervention in operations.
	23	Other	Overall, the government has a limited role, but it makes important interventions through infrastructure upgrades and ongoing operational subsidies. Yet there is something of a gap between the strategic infrastructure investment and the ad hoc FFG/MSRS funding. Likewise, there is a break between private sector operations and public sector planning.

Such freight handling facilities can also be used for retailers stripping containers and consolidating loads for regional stores. Consolidation centres should, thus, also be considered by public planners as a way to support intermodal transport growth.

The findings from this case suggest that the role of the individual decision maker within an organisation could be a subject for future research. In the interviews, it was unclear to what degree a company's interest in using rail is due to a shift in the sector or a purposeful management policy or whether it is just down to an individual in a company. It is, therefore, difficult to drive this through policy, due to the importance of meetings and discussions between 3PL or rail personnel and the potential client, built on individual relationships.

Several future drivers of rail growth are known, including fuel price rises, carbon targets and increasing road congestion, particularly in areas where the road is poor. While the green agenda may have fallen in prominence due to the recession, it remains a key driver, according to interviewees. Fuel price will remain an issue; some operators already update their costs on their contract weekly due to changing fuel costs. Congestion is less urgent at the moment but is an ongoing concern, and corporate social responsibility has grown in importance, according at least to company reports and promotional literature (Jones et al., 2005).

Many of the operational issues identified in this chapter can be observed in other industry sectors, so these findings can be applied in other market contexts. Many observed issues have no obvious solution, such as short distances, fragmented flows, backhaul sourcing, reluctance to share trains, container imbalances and the lack of daytime paths limiting lead times and asset utilisation. Moreover, ongoing public subsidy could be removed at any point. It could be concluded that even the most successful users of intermodal transport have made only small advances towards solving the perennial problems identified in the literature. Intermodal transport is unlikely to grow until the issues identified in this chapter have been resolved, meaning that intermodal corridors cannot yet be the means of control for ports that port regionalisation implies.

Lessons from this chapter that can be applied in other contexts arise from the approach taken in this research, which was to include retailers, 3PLs and rail operators as they all combine to produce successful retail intermodal logistics. Rail service provision in the UK is competitive but as one provider has become more experienced and built better relationships with retailers and 3PLs, this one provider now dominates, and entering this market will be difficult for others. Only one large retailer is directly involved in intermodal logistics, while the others only participate through the use of 3PLs. The 3PL is, therefore, the main player in retail intermodal logistics in the UK, with competition between three providers, all of which have been successful in attracting and aggregating small flows.

The above findings suggest that, along with consolidation centres and multi-user rail terminals, the role of flow aggregators such as 3PLs should be considered by public planners in their actions to support intermodal transport. The approach taken by government policy, planning and funding to facilitate private sidings to support retailer preference does not sufficiently incorporate the role of the 3PLs, which require the prioritisation of consolidation centres and common-user terminals, supporting the aggregation that underpins the financial viability of rail transport. Future research should address this problem, and these findings can also be translated to other market and geographical contexts as well as informing government approaches to the support of intermodal transport.

The case confirmed the literature that government grants have been instrumental in supporting the shift from road to rail, although some concerns exist that the FFG system could have been used more strategically. Generally, however, the role of government was considered more in terms of infrastructure upgrades than direct intervention in operations. A finding from this case warranting further consideration is that concern was expressed by some interviewees that planning consent for more single-user sites could split scale economies and make rail more difficult, although the problem is that shippers prefer their own connections rather than sharing either trains or terminals.

Conclusion

Before discussing the relevance for port regionalisation, some findings can be drawn from the specific case analysis. Findings show that only one retailer in the UK is large enough to fill trains, and that only one rail operator runs almost all of the services. Operational issues in the UK were identified, such as asset utilisation, lack of daytime paths and the crucial role played by wagon and container management. The acute equipment imbalance on the Anglo-Scottish route remains unresolved, and government funding continues to underpin intermodal services. Neither horizontal nor vertical integration are occurring, although some vertical collaboration is clear between Tesco, Stobart and DRS. Private sidings continue to be preferred over common-user terminals, which splits economies of scale and can be a barrier to greater use of intermodal transport. Centralisation tendencies remain strong, and other trends such as port-centric logistics face challenges from the centralised inland system. Finally, difficulties in driving intermodal transport through government policy have been identified, but natural growth may come from drivers such as fuel price rises, carbon targets and increasing road congestion.

Considering the role of logistics integration and inland freight circulation in port regionalisation, this chapter has shown how operational constraints and spatial development of markets can limit the development of intermodal transport, thus challenging the success of inland terminals and rail corridors to ports. In particular, the economic feasibility of these links can be threatened by operational limitations and industry inertia. Even with ongoing government subsidy, it remains difficult to compete with incumbent road hauliers. There is little evidence of major inland market players having an interest in collaboration with ports. Maritime and inland flows remain quite separate, at least as far as large shippers are concerned.

The case study has shown that the 'increased focus of market players on logistics integration' (Notteboom and Rodrigue, 2005: p.301) is not as integrated as the port regionalisation concept tends to suggest. Large retailers are seeking greater control of their distribution in some ways, but it was seen in the UK that only one retailer is large enough to employ factory gate pricing. Furthermore, all retailers studied work in partnership with 3PLs but they are not integrating with them in the way that 3PLs have integrated to some degree with road haulage. Again, 3PLs work closely with rail operators but are not integrated with them. So a series of complex relationships still persists in the industry, suggesting that the dominant players in the market for inland freight circulation are not integrating or even cooperating to the extent that shipping lines are for maritime flows.

The case shows that horizontal integration is not occurring, nor is vertical integration. Reluctance for such integration or even collaboration is a barrier to consolidation, which is necessary for greater use of intermodal transport. If this does not happen, then the ability of ports to capture and control hinterlands through intermodal corridors will remain challenged.

Spatial development patterns based on logistics rather than transport prevent port regionalisation from occurring. While this can be inferred from previous work in the literature, the specific case has shown a lack of interest of market players in integrating with logistics or transport players, which also presents a barrier to the kind of integration with ports required for true port regionalisation processes to occur. Furthermore, the disjunction between different wagon and container configurations required for port and inland flows (a problem in other countries as well as the UK) challenges the economic feasibility of intermodal corridors, further hampering any ability of ports to dominate such corridors. Port regionalisation cannot take place while the two systems remain separated by such operational issues, and little incentive has been identified for any user to solve this collective action problem. Additionally, the corridors remain common-user and competitive and do not offer opportunities for ports to control their hinterlands. The public sector has some influence, but this is limited due to the privatisation and fragmentation of the industry.

Some ports are pursuing direct intervention by promoting the concept of port-centric logistics, which may indeed be part of a regionalisation strategy. However, it has been shown how centralisation is difficult to overcome in the UK because DCs are embedded in these networks. Most importantly, they will not alter their logistics to suit their transport, and most DCs remain non rail-connected.

While rail remains a marginal business, while the industry remains fragmented, while consolidation is not happening and fragile government subsidy is still the basis of many flows, intermodal corridors cannot be the instruments of hinterland capture and control for UK ports. These findings can be transposed to continental Europe, where similar issues persist. The analysis of market and operations shows that even if they want to, ports do not currently have the opportunity to use these links in a competitive way because the industry itself has not solved its own problems. Thus a detailed case study on market and operations has elucidated some issues that prevent the kind of processes that the port regionalisation concept suggests are happening.

Chapter 6
Case Study (USA): Intermodal Corridor

Introduction

This chapter examines the development of an intermodal corridor in the United States, offering the opportunity to study a collective action problem in detail, in which several actors come together to solve a joint problem. Collective action is an arena where various actors can be influential due to the role of informal networking in managing freight corridors; however, institutional constraints such as a conflict between legitimacy and agency and the limitations of institutional design restrict their ability to act directly. The case study demonstrates how the institutional setting of intermodal corridors is changing through the influence of public-private partnerships on government policy. A reconciliation is identified between top-down planning approaches and bottom-up market-led approaches.

A Review of the Literature on Collective Action and Intermodal Corridors

Infrastructure, Operations and Institutions

As discussed in Chapter 2, a corridor can be defined as an accumulation of flows and infrastructure (Rodrigue, 2004). In some ways the concept of a corridor is somewhat arbitrary and may be used for branding or public relations purposes. This is because, beyond a specific piece of infrastructure (e.g. one road or rail line between two places), a corridor usually denotes a large swathe of land through which multiple routes are possible along numerous separate pieces of infrastructure with many different flows organised and executed by different actors. Rodrigue (2004) distinguished between the physical transport infrastructure, the articulation points (e.g. terminals and distribution centres) and the cargo flows distributed along a corridor. This section considers different ways such aspects of intermodal corridors can be analysed.

As with intermodal terminals (see chapter 4), intermodal corridors are frequently analysed by the key performance indicators of operational use, such as cost, time, emissions, capacity, etc. Regmi and Hanaoka (2012) took a holistic approach in comparing two long-distance corridors linking north-east and central Asia. They analysed infrastructure provision, capacity and condition and assessed the time/cost of transport along each corridor as well as administrative processing at border crossings. Other studies have modelled transport times and costs on intermodal corridors, such as Beresford (1999), Janic (2007), Kreutzberger (2008),

Frémont and Franc (2010) and Kim and Wee (2011). Numerous studies, as mentioned in Chapter 4, have modelled intermodal costs more generally, whether that be intermodal terminal location or hub-and-spoke network design.

Challenges in competing with road haulage on cost and time have been identified in the above studies, due to the complexity of intermodal transport, the transloading and inevitable delays, as well the difficulties of bundling flows to maximise capacity utilisation. The advantage of a regularly-serviced high capacity corridor is that it can overcome these difficulties. However, in order to achieve that goal, many organisational and institutional difficulties have been encountered. Some studies have examined these aspects of corridors, involving the numerous transport and logistics operators acting within various governance frameworks and government funding, regulation, policy and planning regimes. Rodrigue (2004) demonstrated how freight corridors represent the regional scale of freight distribution, linking the local and global levels. Thus the globalised distribution channels underpinning the physical separation of freight production, manufacture and consumption connect local zones of production and consumption through regional corridors.

A corridor can also be defined and hence analysed from a spatial perspective. A spatial analysis could investigate both the actual infrastructure provision as well as the density of land use along a corridor length, which leads to related issues such as employment, growth in residential buildings and so on. It can also include links to the corridor and access to and from it, for example the transport, congestion and noise effects on areas around transport and logistics nodes and hubs. Thus spatial analyses lead quickly into political and institutional issues.

The corridor focus is attractive as a planning concept, but, as international corridors cross national and regional boundaries, they do not always fit comfortably into national spatial plans and at regional and local levels there is fear of 'ribbon development' (Priemus and Zonneveld, 2003) or uneven economic development that simultaneously benefits some areas while depleting others (Chapman et al., 2003). There is concern that the space of flows takes precedence and hence replaces the space of places (Albrechts and Coppens, 2003, drawing on Castells, 1996). In resolving these issues, a move from government to governance and particularly multi-scaled governance (Romein et al., 2003) is essential in managing such top-down and bottom-up perspectives. The spatial and institutional complexity of the collective action problem represented by corridor development results in conflicts between top-down command-and-control approaches and bottom-up market-led perspectives (de Vries and Priemus, 2003).

Lehtinen and Bask (2012) showed the difference the correct business model can make in successful corridor development. Similarly, branding of a corridor, with a specified institutional structure and governance regime, can be important elements of corridor success, whether for attracting funding, resolving operational problems or harmonising regulations at border crossings (Kunaka, 2013). Such institutional initiatives have been tried in Africa for long distance corridors crossing several countries, some of which are land-locked and have a significant interest in developing efficient transport corridors for access to neighbouring ports (Adzigbey et al., 2007).

It is thus shown how for intermodal corridors, as with terminals and logistics platforms, spatial development is to a large degree an institutional problem, as intermodal corridors involve many actors that are integrated at different levels and managed by varying arrangements. This finding fits within the port regionalisation discussion in which collective action was derived as the third key element to be examined, and forms the subject of this case study chapter.

Collective Action

Panayides (2002) discussed the costs and benefits arising from different governance structures of intermodal transport chains, showing how the various characteristics of intermodal transport result in a variety of transaction costs. This approach has since been followed in other research on intermodal chains, making use of theory from institutional economics (e.g. Coase, 1937; Williamson, 1975, 1985; North, 1990; Aoki, 2007) to analyse cooperative behaviour in intermodal transport corridors.

Van der Horst and de Langen (2008) highlighted five reasons why coordination problems exist: unequal distribution of costs and benefits (free rider problem), lack of resources or willingness to invest, strategic considerations, lack of a dominant firm, risk-averse behaviour/short-term focus. A variety of coordination mechanisms have arisen to manage this process, such as vertical integration, partnerships, collective action and changing the incentive structure of contracts (Van der Horst and De Langen, 2008; Ducruet and Van der Horst, 2009; Van der Horst and Van der Lugt, 2011).

De Langen and Chouly (2004) proposed the concept of the hinterland access regime (HAR), in which hinterland access was framed as a governance issue because individual firms face a collective action problem: 'Even though collective action is in the interest of all the firms in the port cluster, it does not arise spontaneously' (p.362). The authors defined the hinterland access regime as 'the set of collaborative initiatives, taken by the relevant actors in the port cluster with the aim to improve the quality of the hinterland access' (p.363). Six modes of cooperation were identified: 'markets, corporate hierarchies (firms), interfirm alliances (joint ventures), associations, public-private partnerships and public organisations' (p.363). Five factors influencing the quality of the HAR were the presence of an infrastructure for collective action, the role of public organisations, the voice of firms, a sense of community and the involvement of leader firms. The HAR framework was used by de Langen (2004) and de Langen and Visser (2005) to analyse collective action problems in port clusters.

The study of collective action problems fits within new institutional economics; however the five HAR indicators enable an analysis of the effects of space and scale and thus take a more geographical approach. Relating this framework to the institutional literature reviewed in Chapter 3, these five indicators have much in common with the four indicators of institutional thickness. Institutional thickness is a measure of the institutional setting, while hinterland access regimes refer to specific projects. The aim of the theoretical framework used in this chapter[1] is to draw both approaches together.

Groenewegen and de Jong (2008) applied a new institutional economics model (derived from Williamson, 1975, and Aoki, 2007) to an analysis of institutional change in road authorities in the Nordic countries. The conclusion from their analysis was that those models were unable to capture the complexity of political power play and social and cognitive learning among actors, so the authors developed a ten-step model through which actors become 'institutional entrepreneurs'. Actors benchmark their own 'institutional equilibrium' against a new 'pool of ideas', then spread this new belief system through 'windows of opportunity', using their own 'power instruments or resources', also dealing with 'reactive moves made by the formerly dominant actors' (pp.68–9). While ostensibly working in the field of institutional economics, their approach fits well into earlier discussions of agency and legitimacy found in sociological institutionalism. Aoki (2007) also contributed interesting ideas in relation to how a political champion can alter the game.

Developing the Research Factors

The research factors adopted for this research result from a combination of institutional thickness (discussed in more detail in the institutional literature review in Chapter 3) and hinterland access regimes, modified to include insights from MacLeod (1997; 2001) and others on the role of the state, Groenewegen and de Jong (2008) on actor behaviour game theory and Van der Horst and de Langen (2008) on defining the collection action problem. The research factors are as follows:

1. Reasons for the collective action problem;
2. The institutional setting 1: the roles, scales and institutional presence of public organisations;
3. The institutional setting 2: the presence (or otherwise) of a well-defined infrastructure for collective action;
4. The kinds of interaction among (public and private) organisations and institutional presences;
5. A common sense of purpose and shared agenda;
6. The role of leader firms.

---

1   The six-point research framework used in this chapter was first developed in Monios and Lambert (2013a).

Case Study Selection and Design[2]

The aim of this chapter was an in-depth analysis of solving a collective action problem. This chapter is based on a case study of the Heartland Intermodal Corridor, linking the Port of Virginia to Columbus Ohio, and eventually to Chicago. This was the first multi-state public-private rail corridor project in the United States. Peripheral regions along the route such as West Virginia hoped to decrease transport costs and increase competitiveness by upgrading existing branch lines to double-stack capacity and building new intermodal terminals to access the line.

The field work for this case study took place during 2010. The primary interviewees were those along the route of the corridor, whether directly involved in the planning and development of the project or those benefitting from or otherwise using or related to the corridor. Additional interviews were arranged with senior planners in the federal DOT, which provided additional insights into the planning process, as did interviews at a selection of inland terminals and ports such as Chicago, Memphis and Long Beach.

The case study is presented as a single case narrative. The data collection was guided by the research factors, based on the literature review. The analysis in this chapter is based on a matrix, which breaks down the six factors of the framework into 27 sub-factors, clarifying how inferences were drawn and conclusions reached.

Presenting the Case Study

Freight Transport in the United States

Table 6.1 (below) lists the top ten US container ports in 2009. The dominance of the Los Angeles/Long Beach port complex (known collectively as the San Pedro Bay ports) can clearly be seen. However, two challenges can be identified.

The first challenge to west coast dominance is the expansion of the Panama Canal, from the current Panamax limit of around 4,500–5,000 TEU (depending on design – see Van Ham and Rijsenbrij, 2012) to accommodating 13,000 TEU vessels by 2014. This expansion will mean that large vessels coming from the Far East to bring cargo for the eastern United States can transit the canal and steam directly into east coast or gulf ports (draft permitting). The port of Virginia at Hampton Roads is likely to be the primary beneficiary of this development. New York/New Jersey has the requisite draft, but is currently limited by air draft restrictions (although plans are underway to raise the Bayonne Bridge). Other ports in the Gulf and the Atlantic are also struggling to get the necessary depth to accommodate these larger vessels. On the other hand, the additional time taken to

---

2   This chapter draws extensively on a case study previously published in Monios and Lambert (2013a).

Table 6.1   Top ten US ports by container throughput in 2009

USA ranking	World ranking	Port name	Trade region	Total TEU
1	15	Los Angeles	West Coast	7,261,539
2	18	Long Beach	West Coast	5,067,597
3	20	New York/New Jersey	East Coast	4,561,831
4	41	Savannah	East Coast	2,356,512
5	51	Oakland	West Coast	2,051,442
6	58	Houston	Gulf Coast	1,797,198
7	59	Norfolk	East Coast	1,745,228
8	63	Seattle	West Coast	1,584,596
9	65	Tacoma	West Coast	1,545,855
10	74	Charleston	East Coast	1,277,760

Source: Author, based on Containerisation International (2012)

traverse the canal and reach the east coast may be unattractive to shipping lines. For example, to reach Chicago via Los Angeles/Long Beach takes 14 days at sea plus five days on rail, compared to approximately 25 days to reach Norfolk by sea from Shanghai, with an additional two days to Chicago.

The role played by the Los Angeles area as the largest manufacturing centre in the country suggests that many forwarders will not be prepared to forego the economies of scale that can be gained by transporting all US cargo to this location then separating freight for inland destinations at this point for onward transportation by rail. An additional challenge to the Panama Canal derives from other influences on liner network design. In March 2013, Maersk began serving the Asia-East Coast US route with 9,000 TEU vessels via Suez rather than with 4,500 TEU vessels via Panama. This not only avoided the Panama Canal restrictions and therefore enabled cost savings per container through the use of larger vessels, but also provided an outlet for excess tonnage cascading down from Asia-Europe trades due to the arrival of larger vessels coming into service (Porter, 2013). The gradual westward movement of some manufacturing in the Far East (to India, Thailand, etc.) increases the attractiveness of the Suez route.

The second challenge is the port of Prince Rupert in Canada, which provides a one-day shorter west coast option to shipping lines seeking to access North American markets. This port offers current capacity of 500,000 TEU (with the possibility of expansion up to 2m TEU), in addition to sufficient depth to accommodate container ships up to 12,000 TEU (Fan et al., 2009). In 2009 the port handled 265,258 TEU (Containerisation International, 2010).

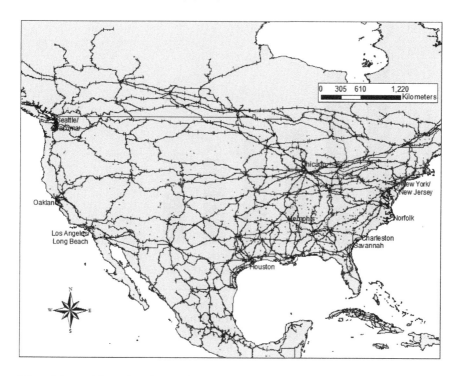

Map 6.1   Map showing major ports and inland locations in the USA
Source: Author

It is unlikely that either of these threats to the dominance of the San Pedro Bay ports will capture more than a small percentage of their cargo. However, as will be discussed below, the hinterland access strategies of these ports will have a determinative impact on port competition.

Map 6.1 depicts the major ports and inland terminals in the USA. There are three classes of railroads in North America: Class I (national), II (regional) and III (shortline).[3] Not including passenger railroads (Amtrak in the US and Via Rail in Canada), there are currently nine class I railroads (annual revenues in 2008 of over $401.4 million) operating in North America. Seven operate in the USA: the big four (BNSF and UP in the west, CSX and NS in the east) plus the two Canadians (CN and CP) and the smaller KCS.[4] There are also two in Mexico: Ferromex and Kansas City Southern de México (wholly owned by Kansas City Southern).

---

3   There is also a fourth class of railroad that performs switching and terminal operations.

4   For ease of reference, the following abbreviations are used. BNSF: Burlington Northern Santa Fe, UP: Union Pacific, NS: Norfolk Southern, CN: Canadian National, CP: Canadian Pacific, KCS: Kansas City Southern. Note that CSX is the full name and not an abbreviation.

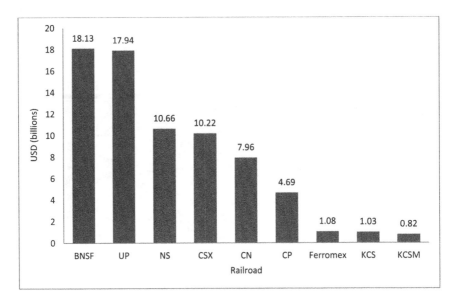

Figure 6.1   Annual revenues at class I railroads in 2009
Source: Author, based on AAR (2010)

2009 revenues at each of the North American Class I Railroads are shown in Figure 6.1. The difference in revenue between western (BNSF and UP) and eastern (NS and CSX) US railroads is striking, illustrating the earning potential arising from greater distances and fewer interchanges characteristic of the western United States railroad network.

A number of key differences exist between rail operations in the USA and Europe, which will now be addressed in turn.

Rail has a higher market share in the US than in Europe because its ability to generate economies of scale makes it the natural mode for long distance hauls. This is most notable for the western railroads which enjoy longer distances and fewer interchanges than the eastern operators. Double-stack capacity on many lines, in addition to train lengths of over 10,000ft in some cases mean that US trains can reach capacities of 650 TEU (compared to around 80–90 TEU in Europe, where bridge and tunnel sizes restrict the ability to double stack containers in most instances and train lengths are also limited by the track infrastructure). Therefore Class I railroads are profitable businesses, unused to government intervention.

US railroads are vertically integrated, which means that each company owns its own tracks and rolling stock and in most cases terminals. They operate completely separately from one another, although railroads may allow for track usage in certain situations. In Europe, rail operating companies compete with each other on common-user track, with the result that maintenance issues are often the responsibility of the public sector. Another salient point is that the eastern and western United States railroad networks are entirely separate from each other. BNSF and UP compete from the west coast to Chicago, while NS and CSX compete between Chicago and the east coast. The two Canadian railroads, CN and CP operate predominantly in Canada, although CN runs down from Chicago, through Memphis to the Gulf of Mexico.

All six class I railroads mentioned above meet at Chicago. This confluence of rail networks means that the city is the location of some of the largest intermodal terminals in the world, each handling many hundred thousand lifts per year.[5] The Chicago area includes approximately 900 miles of track and 25 intermodal terminals, accommodating roughly 1300 trains daily (McCrary, 2010). Freight needing to cross Chicago has to be transported between east and west coast railroad terminals, either by rail or road. Complete trains that do not require reworking will change crew and power at the arriving terminal and then depart, but trains carrying containers for more than one destination will need to be split and reassembled into new trains that may then need to be transported to another railroad terminal across town. This reworking process can take up to 48 hours, therefore road hauls or 'rubber tyre transfers' are more common. According to Rodrigue (2008: p.243), 'about 4,000 cross-town transfers are made between rail yards each day averaging 40km each'.

Increasing congestion at grade crossings eventually led to the CREATE (Chicago Region Environmental and Transportation Efficiency Program) project, launched in 2003. CREATE is a public-private partnership involving the federal DOT, the state of Illinois, the city of Chicago, all the Class I railroads (except KCS) and the passenger lines Amtrak and Metra. The project group was formed in order to seek funding for a number of individual engineering works, including: 'six grade separations between passenger and freight railroads to eliminate train interference and associated delay; it includes twenty-five grade separations of highway-rail crossings to reduce motorist delay, and improve safety by eliminating the potential of crossing crashes; and it includes additional rail connections, crossovers, added trackage, and other improvements to expedite passenger and freight train movements nationwide' (FRA, 2012; unpaginated). The entire project cost has been estimated at $1.534bn, $232m of which will come from the railroads, described as 'an amount which reflects the benefits (as determined by the Participating Railroads and agreed to by CDOT [Chicago Department of Transportation] and IDOT [Illinois Department of Transportation] prior to the execution of this Joint Statement) they are expected to receive from the Project' (CREATE, 2005: p.15). The remainder of the funds are expected to be sourced through a variety of federal, state and local sources. $100m was received from a 2005 SAFETEA-LU federal earmark (see below) and another $100m through the TIGER grant scheme in 2010 (see below).

Any analysis of maritime container flows must remain cognisant of the dominant scale of domestic cargo in the USA. As well as the millions of international

---

5 For a detailed case study on freight management in Chicago, see Cidell (2013).

containers, 89 per cent of cargo in the USA is domestic (FHA, 2010). The added complication is that domestic cargo moves in 53ft boxes (as opposed to 40ft and 20ft maritime containers). It therefore makes operational as well as financial sense to transload foreign cargo at or near the port from 40ft containers into domestic 53ft boxes, as it is cheaper per tonne for trucks and trains because fewer boxes are moved. This tendency has the unfortunate consequence of increasing the cost of repositioning empty containers for outbound shipments from many hinterland markets that do not have sufficient inbound international container traffic. On the other hand, as the US is a net importer, taking a maritime container thousands of miles inland without an export load to send back means that the container will need to be shipped back empty to the port. About 25 per cent of all international cargo moved by rail is transloaded into these domestic containers (Rodrigue and Notteboom, 2010). In the area surrounding Los Angeles/ Long Beach, millions of square feet of warehousing are dedicated to these transloading activities.

The Government's Role in Freight Transport in the United States

Intermodal freight transport was progressing in the 1960s and 1970s, but a number of laws passed in the early 1980s encouraged the development of cooperation between different transport organisations. The Staggers Act of 1980 partially deregulated some aspects of the railroad industry. The number of crewpersons needed for each train was reduced, thus lowering total labour costs for each train. Second, the Staggers Act removed several pricing and scheduling limitations, which increased the railroad's flexibility in meeting market needs. The goal of these changes was to make the railroads more competitive for long distance domestic freight that had been lost to road haulage during the 1970s. As in other countries, this eventually led to a number of mergers and there are currently nine Class I railroads operating in the USA (see above). The Shipping Act of 1984 allowed an ocean carrier to provide inland distribution on a single through bill of lading as well as relaxing numerous other restrictions.

The Intermodal Surface Transportation Efficiency Act (ISTEA) (1991) heralded something of an intermodal approach to highway and transit funding, including collaborative planning requirements (Chatterjee and Lakshmanan, 2008). Supplementary powers were given to metropolitan planning organisations (MPOs), and High Priority Corridors were designated in the National Highway System. In 1998, the Transportation Equity Act for the 21st Century (TEA-21) authorised the federal transport programme until 2003. Several planning objectives for regional transport plans were introduced by this act, including safety, economic competitiveness, environmental factors, integration and quality of life. However, despite these attempts to foster an intermodal approach to transport planning, key government agencies (such as DOT departments) and industry bodies remain modally-based (Holguin-Veras et al., 2008).

Two key issues influencing discussions on future legislative efforts include the Jones Act (1920) which requires that any vessel operating between two US ports must be US-built, -owned, -registered and -manned, and the Harbour Maintenance Tax (HMT). Introduced in 1986, the HMT is a federal tax imposed on shippers based on the value of the goods shipped through ports. Its purpose is to fund maintenance and dredging of waterways, which are the responsibility of the US Army Corps of Engineers. Perakis and Denisis (2008) discussed the obstacle that HMT presents to the development of short sea shipping in the US. As the tax is applied at every port, a water leg in an intermodal chain will attract this fee, whereas transloading to road or rail will not. In addition, a national ports policy is not possible because the US Constitution places limits on the role of the federal government in relation to ports (Talley, 2009).

US federal freight policy is now moving towards a more integrated transportation system; however, the necessary funding to do so remains in various agencies that do not necessarily have the authority to work on cross-jurisdictional projects. Modal agencies are typically responsible for safety rather than infrastructure investment, while in other cases the role of regulatory oversight may be based in a different agency altogether. Finally, the infrastructure for waterways is actually managed by the US Army Corps of Engineers, and not the federal Department of Transportation.

The federal DOT allocates money to the state DOTs and they decide how to spend it, providing little incentive for states to spend money on projects that are perceived to be of primary benefit to other states. Yet, there appears to have been a realisation at national level that cross-border projects should be promoted, which led to the projects of national and regional significance (see below). These represented a noteworthy change of direction, in particular because railroads were eligible rather than just road projects.

The Safe, Accountable, Flexible and Efficient Transportation Equity Act: A Legacy for Users (SAFETEA-LU) (2005) provided approximately $1.8bn in congressional earmarked funds for designated projects of national and regional significance. These funds were allocated to large infrastructure projects decided by Congress and generally driven by politicians on behalf of their constituents, a point subject to some criticism (Proost et al., 2011). Projected benefits could include improving economic productivity, facilitating international trade, relieving congestion, and improving safety. Examples included the CREATE project (see above) and the Heartland Corridor, the subject of this chapter.

The American Recovery and Reinvestment Act (2009 – also referred to as the stimulus package), provided $1.5bn for transport projects through the Transportation Investment Generating Economic Recovery (TIGER) programme. The funds were available for all transportation projects (not just freight) and would be awarded on a competitive basis, with applications due in September 2009 and announcements made in February 2010. The process in TIGER applications was for private money to be matched by public money. The five major goals for TIGER grants were economic competitiveness, safety, state of good repair, liveability and environmental sustainability. This was the first time money was awarded via this method, and a second round of $600m was awarded in September 2010.

The programme proved to be hugely popular; the DOT received almost 1,500 applications totalling nearly $60bn for the first round, and almost 1,000 applications totalling $19bn for the second round. This level of applications suggests the unmet need for such a program, but also reveals the difficulty for the DOT to assess so many divergent applications.

As a result of this round of funding, a significant revival of interest in rail projects can be identified. As the only eligible applicants were public bodies (e.g. states, ports, MPOs), Class I railroads were required to form partnerships with them in order to process an application. An interviewee from the Federal Railroad Administration said that the list of recipients indicates that the assessors of the applications favoured those projects that took an integrated approach to transport problems by focusing on corridors.

The majority of awards were for public transit programmes, highways and other infrastructure upgrading, but freight-specific projects, such as transportation hubs and port upgrades, also received financial support. Some larger freight projects included $98m for the CSX National Gateway project (see below), $100m for the Chicago CREATE project (see above) and $105m for the Norfolk Southern Crescent Corridor (see below). A marine highways project in California was also among the recipients.

The Energy Independence and Security Act (2007) included a provision for 'America's Marine Highway Program' to integrate the nation's coastal and inland waterways into the surface transportation system. Therefore Congress instructed the DOT Maritime Administration (MarAd) to create a Marine Highway program that examined ways to utilise waterways where they may provide some services on parallel highway routes to alleviate bottlenecks. They assessed the country's waterways and invited applications and in August 2010 eventually designated 18 marine corridors, eight projects, and six initiatives for further development. These eight projects were then eligible to bid for a total of $7m in pump-priming funding to develop projects.

An increasing focus on infrastructure corridors can be identified in this case study. These projects incorporate several localities, regions and states, not to mention private and public stakeholders, and have therefore demanded new methods of management and innovative funding schemes. From a practical perspective, the real role for this money is to enable large consortia to come together where public and private benefits can be clearly identified among all the parties. The federal government could never realistically spend enough money to exert significant influence on the operations of the rail industry, nor would it desire to do so. Indeed, it is not inaccurate to suggest that railroads have in the past been reluctant to accept public money for fear of external influence on their business.

The Heartland Corridor

The Appalachian region follows the Appalachian mountain range in the eastern United States, running from the state of New York southwest to Alabama and

Mississippi. It covers an area twice the size of Great Britain but with only about one-third the population. The Appalachian Regional Commission (ARC) was created in 1965 to coordinate economic development opportunities in this somewhat inaccessible and hence economically disadvantaged region.

The project began in 1999, when the ARC commissioned the Nick J. Rahall Appalachian Transportation Institute (RTI) at Marshall University to undertake studies of commodity flows (phase I) and transport costs (phase II). The results indicated that impediments to shippers in the region were due to poor access to major rail and port traffic routes. Low traffic volumes and difficulties in sourcing backhauls were the operational problems. From an infrastructural perspective, it was the fact that the numerous tunnels necessitated by the rugged terrain had not been built to accommodate double-stack clearance. These issues all resulted in increased trade costs to local shippers (RTI, 2000). The recommendation was for a detailed study of new track and terminal infrastructure, including costs and benefits for all parties such as region, state and national governments, as well as the railroad companies.

This second study was eventually commissioned following initial meetings in 2000–2001 between the steering group, which comprised West Virginia DOT, Norfolk Southern railroad, the Ohio Rail Development Authority, Virginia Department of Rail and Public Transportation, ARC and RTI (ARC, 2010). The funders of the research were the states of West Virginia and Ohio, RTI, Norfolk Southern and the Federal Highways Administration (FHA), and the brief was to select the best route, estimate the costs of double-stack clearance and measure the project benefits for all stakeholders. Initially, both eastern railroads (Norfolk Southern and CSX) were invited to take part, but after early meetings CSX declined therefore only Norfolk Southern routes were considered. The route selected by the study (see Map 6.2 below) was used primarily by coal trains delivering coal to the port of Virginia, location of the largest export coal terminal in North America. The inability of this line to accept double-stacked container trains limited the competitiveness of regional shippers, who thus had to drive longer distances to access intermodal terminals. According to the study, a penalty of $450–650 applied to each container movement.

For the lower Appalachian area, the main container routes were through west coast ports then by rail via the Chicago hub (where the container will change from western to eastern railroads), or via east coast ports. If using an east coast port such as Virginia, the rail options were single-stack direct or double-stack on a route that added over 200 miles and around 24 hours to the journey, hence increased cost. The conclusion was, therefore, that trade to this region was penalised. At the time, West Virginia was ranked 40th out of 50 states in the percentage of Gross State Product derived from exports (RTI, 2003).

The high costs of engineering work (initially estimated at up to $111m) resulted from the challenging topography in this mountainous area, but the Benefit-Cost-Ratio was estimated at between 2.0 and 5.1. The American Association of State Highway Transportation Officials (AASHTO) identified the corridor as one of

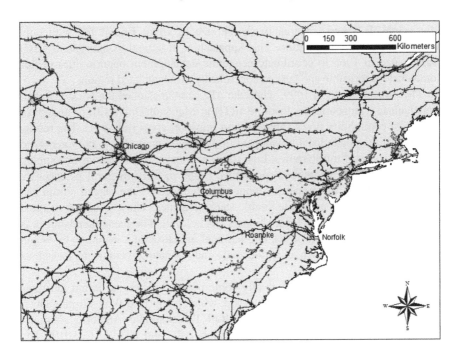

Map 6.2   Map showing the Heartland Corridor route with inland terminals
Source: Author

three multi-state rail projects with the potential to deliver considerable public benefits (AASHTO, 2003). Despite substantial benefits for the private sector, the study concluded that the corridor upgrade was unlikely to proceed without public support. The project therefore required the steering group to raise considerable public interest: 'In state capitals, town halls and business offices, and on Capitol Hill, scores of familiar questions were answered thousands of times, while both methods and conclusions were constantly scrutinized. As a product of this process, untold numbers of stakeholders helped shape and promote what eventually became a coherent legislative initiative' (ARC, 2010: p.9).

According to local interviewees, these public meetings helped the representatives at the federal level realise the importance of developing corridors of national significance. Several meetings were held in Washington DC to promote this agenda and the interviewees agreed that it was the trade argument that interested federal legislators the most. While investing government money in the project would benefit a private company, significant economic development benefits were expected to result from the improved trade access for the region.

The Heartland Corridor was designated as a Project of National and Regional Significance in the 2005 SAFETEA-LU legislation, authorising $95m in federal funds for the project (reduced to $84.4m by estimated obligation limitations, rescissions, etc.). Out of the total cost of $195.2m, $84.4m was federally funded, $101.0 million was contributed by Norfolk Southern, $0.8m from the Ohio Rail Development Commission (ORDC) and $9.0m came from the Virginia Department of Rail and Public Transportation (VDRPT).

Two Memoranda of Agreement were completed in August 2006. One was between the Federal Highway Administration (FHA), the Eastern Federal Lands Highway Division (EFLHD) and Norfolk Southern, and the other between FHA, EFLHD and the three states. The agreements were necessary to identify and agree the roles and responsibilities for the environmental planning, design and construction of the corridor. The first MoA established a funding mechanism that allowed money to flow directly from the federal government to the railroad, unprecedented at the time. As the majority of the tunnels were in West Virginia, the second MoA was required for the other states to agree that the bulk of the money would be spent there. In total, 28 tunnels and 26 other overhead obstructions needed to be raised along the route to allow the passage of double-stacked container trains. Construction began in October 2007 and the first double-stack train ran on 9th September 2010.

Part of the upgraded corridor project was the requirement for intermodal terminals. A new intermodal terminal was built at Rickenbacker, outside Columbus, situated within a large logistics park, named the Rickenbacker Inland Port. The site is owned and operated (through a third party contractor) by rail operator Norfolk Southern, and was built to serve as an alternative hub to Chicago in the northeast of the United States. This integration of a rail terminal with a logistics park was the first time NS had done so (Rodrigue, 2010). The site was developed through a $68.5m partnership between Norfolk Southern and the Columbus Regional Airport Authority (with $30.4m coming from a SAFETEA-LU earmark). While the port of Virginia at Hampton Roads played a supportive role throughout the process, as it will benefit from improved inland access in order to compete with west coast ports, it was not actively involved in the terminal development. The total site covers 1,576 acres. In addition to the common-user terminal, customers can have direct rail connections to private warehouses.

This terminal currently has three trains in and three out per day, and is about to commence a service from Columbus to Virginia. There are also two trains from the east coast to Chicago that get merged into one here, and a domestic train each way to Chicago. The terminal is currently doing 500 lifts per day, but it could do double the current lifts. Unlike many inland ports in the United States, 95 per cent of its traffic is international, coming from the port. The trains will be made up primarily of cars that take 20ft, 40ft and 45ft boxes, rather than 53ft domestic boxes (Boyd, 2010), because the aim of the connection is mainly maritime access. This operational strategy is also attractive because the US network is currently experiencing a shortage of 53ft well cars for double-stack services.

While government attention tends to focus on the infrastructure for the trunk haul, each container needs to come from a distribution centre or warehouse. Each origin or destination may only be contributing a handful of containers a day or week to the terminal, so a large amount of space for logistics operations is required to feed the terminal and make the rail operations economically viable. A large purpose-built site like Rickenbacker, with a good amount of greenfield space for future development, is able to attract companies like 3PLs to locate in one area, thus reducing the road haul and increasing the attraction for intermodal transport (this point will be expanded in the discussion in Chapter 7). According to one interviewee, his firm made an early decision to build a new warehouse at the Rickenbacker site because it was felt that the new rail connection with east coast ports would be 'one of the most important developments for trade in Columbus since the interstate'. Smaller terminals on the route are also planned at Prichard, WV and Roanoke, VA, but funding is still being sourced before work can begin at these sites (see Map 6.2). This is particularly important for shippers in West Virginia who are currently more than 1130 miles from the nearest intermodal terminal, resulting in additional costs estimated at around $450–650 per container (RTI, 2003).

Recent Development of Intermodal Corridors in the USA

The Alameda Corridor was the first major public-private partnership (PPP) intermodal corridor in the United States. It is a short (20 miles) high capacity (three double-stack tracks) line designed to reduce congestion and other negative externalities associated with the extremely high container flows of the San Pedro Bay ports. The corridor involved consolidating 4 branch lines, which reduced conflicts at 200 grade crossings and included a 10 mile trench. The new line was opened in 2002, with a capacity of about 150 trains per day. The project had a total cost of $2.43bn, split between $1,160m revenue bonds, $400m federal loan (the first of its kind), $394m from the ports of Los Angeles and Long Beach, $347m MTA grants and $130m from other sources (Goodwin, 2010).

Unlike the Heartland Corridor, the ports are directly involved in the project; they are the financial guarantors of the corridor and will lose money if the route is not used and incurs losses (Jacobs, 2007; Callahan et al., 2010). While the two ports are located immediately next to each other and have acted together in this instance, they remain separate institutions, each administered by their respective city's harbour department (Jacobs, 2007). In 2009, of the 11.8m TEU through the ports, 3.4m TEU travelled up the corridor (2.8m TEU using on-dock connections and 0.6m TEU near-dock). 0.7m TEU used off-dock rail, 3.4m TEU used rail after transloading into 53ft domestic containers and 4.3m TEU travelled inland by truck. (Goodwin, 2010)

However, while the corridor solves certain problems for the port, it presents other issues for the two competing rail operators. UP operates a large intermodal terminal (ICTF) about five miles from the port. They are able to transport their containers there by truck, which are then consolidated and sent on block trains up the Alameda Corridor and across the country. By contrast, BNSF's main terminal is at Los Angeles, at the end of the corridor. The result of these terminal locations is that BNSF often drive maritime containers to transloading warehouses where the load is reconfigured into 53ft domestic containers; these are then driven to their LA terminal to be put on a train, thus bypassing the corridor. This situation is a result of location decisions made in the early days of intermodalism, when SP (later bought by UP) bought the ICTF terminal due to their business location. BNSF has small rail sidings near the port as well as an agreement with Maersk to use space within their LA port terminal, whereas UP has, in addition to their ICTF terminal, large rail sidings between the port and ICTF. This situation shows that, when rail corridors are built, it is important to understand issues such as train marshalling that can have major impacts on usage of the mainline.

The Heartland Corridor was the first multi-state PPP intermodal project, and its influence can be seen on subsequent developments of large multi-state intermodal corridor projects. CSX's National Gateway is a similar PPP that also joins Norfolk with Ohio, via a different route that involves 61 double-stack clearances, the construction or expansion of six intermodal terminals and will cost $842m (McCrary, 2010), including $98m in funding from the first TIGER programme. Similarly, Norfolk Southern has proposed a number of projects along what has been branded the Crescent Corridor, a 1,400 mile stretch running between New Orleans and New York. The ambitious goal is to develop a PPP to cover the estimated cost of $2.5bn. The project is far larger than the heartland Corridor, involving 13 states, 11 new or expanded terminals and 300 miles of new track. In February 2010 the project was awarded $105m in TIGER I grants through an application from Pennsylvania DOT and in August 2010 six states submitted applications under the TIGER II programme totalling $109.2m, although none were successful.

One result of these corridor developments will be to transform Ohio into an intermodal hub for the US, with both eastern railroads having major intermodal terminals there. This will allow some traffic to bypass Chicago, thus potentially redrawing the map of intermodal transport in the country.

Analysing the Case Study[6]

Table 6.2 (below) sets out the thematic matrix with the relevant data noted against each factor. The table is followed by a discussion of these findings.

---

6   The analysis table in this chapter is an expanded version of the analysis in Monios and Lambert (2013a).

Table 6.2   Applying the thematic matrix

Factor	No.	Sub-factor	Data
1 : The reasons for the collective action problem	1	Unequal distribution of costs and benefits	Public spending is controlled by each state, but, from a corridor perspective, the benefits accrue to more than one state. It is therefore difficult to coordinate such a project where one state may have to spend more because a lot of the rail line is in their state, while another state may receive a disproportionate benefit. From a private point of view, the costs of investing in an upgraded line would bring benefits for users but not enough revenue for the operator to cover the investment.
	2	Lack of resources or willingness to invest	As in 1 and 4, there was no willingness to invest on the part of the two Class I railroads. Interviewees did not directly state the reasons, but it may be because benefits of doing so were not clear, either from the perspective of economic feasibility of the new traffic, or as a competitive act against the other operator. In the public sector, sufficient financial resources were not available within each state to develop their own rail infrastructure, as a corridor perspective was required. This is related to the following sections on scales of governance. Lack of knowledge of rail in the public sector was also noted by both private and public interviewees, as this is not the traditional purview of DOTs in the United States.
	3	Strategic considerations	As in 4, two major rail operators serve the area. There was no need for one firm to capture a market as a competition mechanism because both firms knew that a large investment was required therefore the result was inertia.
	4	Lack of a dominant firm	The area is served by two Class I railroads, Norfolk Southern and CSX.
	5	Risk-averse behaviour/short-term focus	Related to 1–4, there was no incentive to take a long-term view of the payback for the required investment to upgrade this line. Due to the rationalisation of the Class I rail network in previous decades, the focus is now on high capacity lines rather than peripheral routes. The requirement to provide regular dividends to shareholders was noted by one railroad as a constraint on long-term investment.
	6	Other	According to the regional body and the commissioned studies, there was a lack of understanding in the public sector (i.e. the state DOTs) of the importance of regional access to global markets. This was highlighted in the interviews as an important part of the need for a regional focus rather than a state focus.

7	At which level are institutional presences scaled	Roles were well defined at local, state and federal level, and institutional thickness was strongest at state level. State DOTs were and remain the primary conduit for infrastructure planning and investment.
8	Confused sovereignty, multiple authorities and funding sources	Sovereignty was not confused as roles were well defined (see 7). The problem was that the primary scale was the state level, which lacked ability to act. Other funding sources were known (e.g. federal earmark) but there was no clear process of how to obtain this funding, which therefore required much informal networking. However, for the states to lobby effectively in Washington, it was first necessary to achieve a unified project vision at the local and state level. This was driven primarily by the regional body ARC, supported by some key stakeholders such as a political champion in the person of a WV state senator.
9	Constant changing and re-making of institutions	This was not observed during the project as institutions remained stable. After the project, the changing of federal funding for transport projects has altered the institutional setting. While organisations remain the same, the institutional setting (in the sense of the rules of the game) has changed, and now multi-state PPPs have become a key way to acquire funding for transport infrastructure.
10	Limited government organisations due to political designs can mean that delivery of government policies may be 'hobbled'	Political design of transport agencies means that they don't have a great deal of power for direct intervention. However, their roles are clear therefore new stakeholder groups are able to facilitate their own actions. Moreover, the limited state role allows freedom of action for other groups to access funding directly without involvement in planning bureaucracy.
11	Conflict between legitimacy and efficiency	As in 10, the state level of transport governance is the most legitimate scale, but it has limited ability to promote intermodal transport or direct investment there. Conversely, the less legitimate state body ARC was able to use informal networking to promote the project and turn it into a coherent vision that could be lobbied at federal level. Thus a conflict between legitimacy and efficiency (agency is perhaps a better term here) has been observed.
12	Other	

2: Infrastructure for collective action 1: the roles, scales and institutional presence of public organisations

(continued ...)

Factor	No.	Sub-factor	Data
3: Infrastructure for collective action 2: how the system works	13	The rules of the game	As in 8, the rules of the game were known, thus the limited infrastructure for collective action was known, but it was very limited. The rules were that state DOTs controlled infrastructure spending and that rail was predominantly a private sector area therefore money would not be directed there. Thus the rules of the game had led to the collective action problem.
	14	The current equilibrium outcome, i.e. a shared understanding of how the system works	Initially there was not a well-defined infrastructure. Both the rules (see 13), and the understanding of this system by actors, led to an impasse where no actor was prepared to invest. The shared understanding was that it was up to the private sector to invest, and state DOTs would not invest because benefits to each state were not clear. Therefore a political champion was particularly important at the early stages to connect the stakeholders with funding opportunities, thus attempting to alter the shared understanding of how the system works.
	15	Innovation may be stifled by inappropriate formal structures	As in 8, 10 and 11, the structure of transport governance prevented investment in this area. It is not clear from the data that innovation was 'stifled'; it is perhaps safer to infer that, as in 13 and 14, actors understood the status quo and were not incentivised to change the current system. However, there is no evidence in this case that innovation had been tried and stifled by the current structure of transport governance.
	16	Monitoring may become primarily ceremonial and related to the formal structure rather than to the real activities of the organisations	The monitoring of public agencies or the institutional setting did not come up directly in the interviews. Indirectly, however, the structure of the state DOTs, whereby their funding is based on their state transport plans and strategies rather than a regional approach, could be interpreted as a 'ceremonial' monitoring that focuses on the status quo rather than a deeper analysis of issues for regional shippers that were not being addressed. In particular, state funding is focused primarily on highways as rail is a private business, therefore the 'real activities' of the DOTs could be defined as ensuring market access and good transport connections, which were not being monitored in a system that did not monitor rail because it was not a public responsibility to do so.
	17	Other	The multi-state concept of this project was replicated in other projects seeking federal funds, thus a well-defined infrastructure for collective action can now be observed. It is difficult to assess to what degree this project influenced subsequent projects or in particular the new direction of federal transport funding. A clear trend can be observed, which was discussed with interviewees during the research trip, as the second round of TIGER programme applications were being assessed while interviews were being conducted in the USA during September 2010. However, clear relations between the two phenomena are difficult to identify.

18	What actions were taken	Market studies were conducted on behalf of the regional body ARC to identify options. A steering group was established by the ARC with regular meetings, involving both the public and private sectors. In particular, the early involvement of the private sector railroad operator was essential. Regular promotion was used to convince relevant people of the benefits of the project. A high level of interaction was noted by interviewees as being of central importance. Once a coherent project vision had been established, the project was then lobbied for at federal level for congressional funds.
19	Informal collaboration and influence	As noted above, the regional development agency ARC was able to build informal networks, and as discussed in the notes above regarding formal structures of transport governance and their relation to the lack of an infrastructure for collective action, it was that very lack of agency in the system that both necessitated informal networking but also allowed it to succeed. A political champion was particularly important at the early stages to connect the stakeholders with funding opportunities. It is difficult to map such a process objectively as it is reliant on the interview statements which may not describe the reality accurately, therefore caution must be exercised here. However, the informal nature of the process was raised by many respondents.
20	Other	The informal networking has to an extent been institutionalised in the current TIGER grants, which are application-based, and rely, therefore, on bottom-up consortia approaches based on PPPs.
21	Stakeholders established agreement upon the priority and message necessary to complete the task	Regular steering group meetings were held to establish a joint vision to enable the promotion of the project in different contexts. Essential to this vision was agreement on a multi-state corridor approach, which was based on recognition by all stakeholders of the shared benefit in allocating the funds to specific locations. The group developed common presentation and branding materials that contributed to the coherent project vision and the involvement of a political champion helped to drive this project at federal level.
22	Link between establishing the vision and achieving the outcomes	Many interviewees noted that many meetings were required to establish the vision, and many local presentation and discussion sessions took place. It was noted that without this work, the project would not have progressed to federal level.
23	Other	

4: The kinds of interaction among (public and private) organisations and institutional presences

5: A common sense of purpose and shared agenda

(continued ...)

Factor	No.	Sub-factor	Data
6: The role of leader firms	24	Institutional entrepreneurs benchmarking their own institutional equilibrium against new ideas	It was difficult to establish this factor definitively based on interview responses, and it also relates to 26 which is based more on observed actions. It could be said that the rail operator that got involved in the project (as opposed to their competitor who declined) was trying new ideas to alter their institutional equilibrium, but how far this process was actually understood or planned strategically is difficult to map. This factor could be studied better by action research, thus is perhaps not suitable to this framework.
	25	Use their own resources	Norfolk Southern was involved from early on in the project. It contributed funding and staff time and was noted in the interviews with other actors as being flexible with changing project plans.
	26	Leads to reactive moves by other firms	Other firms followed afterwards in similar multi-state corridor projects, most notably CSX's National Gateway (CSX had declined to take part in the Heartland Corridor).
	27	Other	States are now looking at other related investment along the corridor for intermodal terminal access.

State rescaling issues were not identified in this case, as roles and responsibilities were well defined at all levels. MacLeod (2001) advised the requirement to identify at which scale institutional thickness is strongest; in the United States it is clearly the state level, as evidenced by the state DOTs. Yet for this project the key issue was the relation between the states, as a regional impetus was needed to draw states together. The main actor was a relatively weak regional organisation (ARC) drawing together stronger individual organisations, such as state DOTs and the privately-owned railroad Norfolk Southern. The research identified a lack of motivation on behalf of the railroads to challenge the current situation, requiring the promotion by the ARC and others of a greater understanding of the role of regional access to global markets.

While a trend towards devolution has been identified in many areas of the world (Rodríguez-Pose and Gill, 2003), the primary level of institutional presence with regard to transport in the US has been scaled at the state level for a long time (Haynes et al., 2005). The importance of regional cohesion in devolved governance systems as demonstrated in this case can thus be of relevance to other contexts. Hall and Hesse (2013) have noted the increasing importance of the regional scale even though it has traditionally been an institutionally weak level of governance from a formal perspective. This case has shown that the very informality of the regional scale can be utilised effectively in certain conditions.

The notable feature of the existing system from an institutional perspective was the clear roles for existing organisations within their institutional setting. Unfortunately, the reason for this clarity was that public organisations occupied a small role with little influence over the railroad sector. Railroad development in the United States is planned and funded by the private sector, which has little incentive for potentially risky investment. As a well-defined infrastructure for collective action did not exist, informal arrangements, brand development, political championing and congressional earmarks were therefore required to bring the two sectors together and highlight the potential for both private and public benefits.

The institutional setting in the USA has since been altered by the development of more transparent funding systems, based on a clear bidding process for pre-determined funding sources. Throughout this process, organisational arrangements such as PPPs have been encouraged to develop, which demonstrates that a well-defined infrastructure for collective action, not in existence before the project, has since been established. Somewhat counter-intuitively, a move away from congressional earmarks towards a discretionary system means that legislators will potentially have less influence over strategic planning, as such a system would depend on ad hoc bids (see Moe, 1990). Alternatively, if multiple private and public sector partners are required to form consortia in order to attract federal money, greater strategic cohesion across larger areas becomes more likely than via the usual state-by-state approach.

It was shown above that this project was initiated by the regional development agency, but once other stakeholders joined, the project developed its own identity. Of particular relevance in the early stages were regular meetings, promotional events and a political champion. Crucial for the successful development of such multi-partner, or indeed multi-region or multi-state projects is establishing an agreement among stakeholders that the investment will benefit all locations along the corridor (McCalla, 2009). In this case, in order for the public and private funds to be blended a new framework of agreements had to be developed (the MoA discussed earlier), ensuring that funds were spent in certain ways.

A similar resolution between top-down and bottom-up perspectives was observed in the Alameda Corridor project, as the ACTA was able to purchase the rail lines from the rail operators and obtain an agreement that they would use the new corridor. By contrast, the Alameda Corridor East project is a grade separation project to increase safety in the region by removing at-grade railway crossings (Monios and Lambert, 2013b). Callahan et al. (2010) criticised the project for not utilising the same institutional framework as ACTA, but the conflicting stakeholder motivations precluded such a framework. The grade separations do not provide much advantage for the railroad (as they already have the right of way), beyond the removal of the threat of collisions. Second, because the project does not consolidate freight on a new line, it was not possible to purchase the tracks under a new authority as in the case of ACTA. However, the construction work has been managed under a single joint powers authority which has been successful in utilising the branding of the corridor to focus public money that may have been more difficult to attract for each construction project individually.

The analysis of the Heartland case has shown how institutional thickness is not enough on its own. A similar finding was reported by MacLeod (1997) and Pemberton (2000), both of whom emphasised the role of the central state rather than focusing overly on institutionally thick yet still underperforming regions. Raco (1998) discovered that increased institutional thickness at a local level can also reinforce existing inequalities, therefore attention should be paid not simply to the thickness itself but its ongoing processes. Similarly, Henry and Pinch (2001) showed how a region does not require institutional thickness itself if it has superior access to an already institutionally thick national or international scale. The use of an expanded framework in this chapter builds on these findings by explicating how the successful exploitation of institutional thickness depends on the type of institutional structure in both the wider setting and the specific project itself (Monios and Lambert, 2013a).

As discussed in the literature review, institutional thickness relates to the institutional environment, while hinterland access regimes address specific projects. The aim of the theoretical framework used in this chapter was to draw both approaches together. As such, some indicators relate to the overall system, while others cover the role of actors within this institutional setting.

In the United States, the current institutional setting is constituted by the roles, power and influence of public and private organisations, each manifesting in a variety of ways, through such areas as planning, policy, operations and the marketplace. While government policy in the USA is broadly in favour of intermodal transport, planning for intermodal freight does not include direct intervention, because rail infrastructure and services are owned and operated by the private sector. While planners desire to remain informed of any issues where they can assist, it is not their role to intervene and they do not possess sufficient instruments to influence the situation.

Nevertheless, private sector actors can find it difficult to make investments in transport infrastructure. They therefore depend to a significant extent on public sector support, which generally comes through the planning system. The problem is that when public funding is used to support an infrastructure project, questions of ownership are raised, which explains why the private sector has traditionally been wary of accepting public money. As a rule, private organisations will make decisions based on the profit motive, responding to operational requirements or signals from the market. This approach often produces a lack of long-term investment due to shareholder pressure. As shown in this case, however, a public organisation with little institutional presence had the necessary flexibility to draw together policy-led planning departments and profit-seeking private operators. This consortium was able to address the concerns of local shippers in a situation where neither policy and planning actions, nor the operations and market focus of the private sector, were likely to address their problems if the current institutional setting was not altered.

The way the institutional environment was altered in this instance was to provide access to public money for private operators to accelerate their operational requirements, rather than enforcing a planned system of infrastructure investment from the top down. The role for this use of public money was to enable large consortia to come together where public and private benefits could be clearly identified among all the parties, as per the initial brief of the research study with which the project commenced. It would be risky to conclude that the TIGER programme should replace funding through transportation legislation, and, indeed, there is an important conversation to be had on that issue, but its influence will certainly be felt in future plans.

One key aspect of the Heartland Corridor project was that the private sector rail operator was brought into the project first, before it was taken further to lobby for public money. Prior to this project, much resistance was found in the private sector towards government involvement, and the sectors were not integrated in the sense of long-term strategic planning. The achievement of the ARC to build institutional capacity through stakeholder groups involving the private operator along with shippers, thus forming a shared agenda, was extremely important.

In the United States, institutional presence in terms of transport governance is scaled mostly at the state level, while the regional body does not have much actual power of action. However, it managed to create agency through informal networking. Furthermore, taking a corridor approach across jurisdictions was essential to the success of the Heartland Corridor. In the USA, vertical integration in the rail industry makes investment planning by the rail operator much easier, because they have control over the operations in relation to the infrastructure investment. Thus they know that if a certain amount of money is spent, a certain amount of revenue is likely to result from shippers in these regions. The crucial point about investing in intermodal infrastructure is that demand can be consolidated on key routes to create economies of scale and allow operators to bid on this consolidated traffic, rather than many small flows that could not support such expensive infrastructure and services. The Heartland Corridor demonstrates how taking an integrated corridor approach can overcome these problems.

New developments in transport funding from the federal government in the USA are also interesting, because they suggest the influence of the Heartland project, and how a reconciliation can be achieved between top-down planning approaches and bottom-up market-led approaches, as justified in the funding applications by consortia of public and private actors. Therefore path-dependant transport chains can be disrupted by peripheral regions through the coordination of public and private bodies. The success of multi-state PPPs can lead to governance reform in the way transport planning and public investment are restructured to attract private interest, and bring forward large rail infrastructure projects that otherwise would not be pursued.

Coulson and Ferrario (2007) highlighted the importance of not conflating correlation and causation when a successful project and a strong institutional environment are observed. It is, therefore, not possible to conclude that this project caused a change in policy. Yet a clear trend may be observed from federal loans (e.g. the Alameda Corridor) to grants (through earmarks, as in the Heartland Corridor) to competitive bids (the TIGER grants). Additionally, two major multi-state corridor projects have since commenced, based on PPPs and federal funding from the TIGER programme. Another trend may therefore be observed, towards multi-state projects. It can therefore be concluded that governance of transport infrastructure development has progressed towards a reconciliation between top-down planning approaches and market-driven private sector development.

The institutional literature reviewed in Chapter 3 suggested a conflict between legitimacy and efficiency and a limitation of political organisations due to their design, both of which were confirmed in the findings (although agency may be a more accurate term than efficiency in this context). These issues explain the high incidence of policy churn, lack of agency and, sometimes, lack of communication between the public and private sectors. Also important was the role of informal networking as it can overcome institutional inertia; it is, though, difficult to capture this process, and harder still to attempt to institute it in another setting through policy action.

The framework developed in this chapter attempts to join analyses of individual projects and institutional settings, reconciling institutional economics and institutional approaches from economic geography. It is drawn from a vast institutional literature and is therefore fairly broad at this stage; it requires application in more cases, therefore, in order to test it further and improve its relevance and explanatory power.

Conclusion

Before discussing the relevance for port regionalisation, some findings can be drawn from the specific case analysis. A trend in federal funding from loans to grants to competitive bids has been identified, as well as the potential relevance of the TIGER funding grants in other contexts, in which the PPP approach required in TIGER applications may be a way to reconcile bottom-up and top-down approaches to transport planning and funding. The need for a more strategic approach in the UK was discussed in Chapter 5, and the potential applicability of a similar scheme there, as well as other countries, could be a future research topic. Findings from this research on the importance of regional cohesion (informal or otherwise) in devolved governance systems can also be of relevance to other contexts.

This chapter can also build on the findings from Chapter 4, which compared a number of inland terminal case studies. The Rickenbacker inland terminal could be added as another inland-driven case study, in this case being able to develop good relations with the port, as part of the larger corridor project, thereby reflecting the benefits of port involvement from the beginning, as well as the benefits of taking a corridor approach. By contrast, one of the western railroads interviewed complained that they do not use the Alameda Corridor as much as they otherwise would because they do not have sufficient terminal space nearby to marshal trains. Therefore that project becomes a case of a port authority integrating inland to the extent of building the rail infrastructure, but without the operational integration displayed in the Venlo case where the port terminal operator also operates the rail link with the inland terminal. So this chapter has provided additional cases to develop the classifications from Chapter 4.

Considering the role of collective action problem resolution in port regionalisation, this chapter has shown that conflicts between legitimacy and efficiency or agency and a limitation of political organisations due to their design may account for the high incidence of policy churn, lack of agency and, sometimes, lack of communication between the public and private sectors. Thus institutional design constrains integration between maritime and inland transport systems, suggesting that port regionalisation processes will face challenges developing in the way that the concept assumes. This part of the research goes some way towards describing how 'national, regional and/or local authorities try to direct this process' (Notteboom and Rodrigue, 2005: p.306), and explaining how 'a lack of clear insights into market dynamics could lead to wishful thinking by local governments' (p.307). It is the way public organisations are designed that limits their capacity to engage successfully in such situations; when they do, the conflict between legitimacy and agency limits their effectiveness.

This part of the research also highlights the importance of scale. In all situations, it is important to identify at which level transport governance is scaled, and how public and private organisations interact within and around these scales. In this case, a regional body utilised informal networking to overcome the inertia at state level, where transport governance is scaled in the USA. Comparisons can be drawn with Chapter 5, where national funding exists to promote intermodal transport, but the funding scheme is ad hoc rather than strategic. So the results from this case study analysis demonstrate that when discussing the role of the public sector in port regionalisation, it is first necessary to understand the roles and powers of public bodies in that particular country or region, before forecasts can be made of how private freight stakeholders may act.

Similarly, the case study analysis revealed the importance of leader firms, yet these firms are reluctant to act without the infrastructure for collective action being clearly defined. So again, when predicting or explaining the likely path of regionalisation processes in a particular country or region, these aspects must be understood. In this case, there were two competing rail operators. In the case of the UK retail sector in Chapter 5, there were a number of retailers competing, but only one large enough to fill a train; as the other retailers would also prefer a private train rather than a shared service, this prevents them exploring rail to a greater extent.

More cases are required to validate these suggestions, but the details of how a case works in practice can raise issues about the extent to which port regionalisation can actually happen and what is required for it to happen. The case elucidates good reasons why ports may experience challenges in controlling or capturing hinterlands through the strategies of integration that the port regionalisation concept suggests.

Institutional design constrains, or at least challenges, such regionalisation processes of integration from occurring; therefore, in future research greater disaggregation of port regionalisation possibilities could be pursued along the lines of institutional models of ports and other stakeholders, particularly public sector planners and funders. Legitimacy and agency are a problem for these organisations and if an infrastructure for collective action is not in place (and it is usually predominantly a *public* infrastructure for collective action), then private firms will not act, thus hampering any attempts at port regionalisation and keeping the maritime and inland spaces separate.

Therefore for port regionalisation to happen in a region (at least in terms of intermodal transport which is the limited focus of this research), it will depend on the institutional setting and relative constraints on action, including design of public organisations and the conflict between legitimacy and agency.

Chapter 7
Institutional Challenges to Intermodal Transport and Logistics

Introduction

This chapter returns the findings from the three empirical chapters to the theoretical context of institutional analysis. The key relationships are identified and explored in the context of other cases in the literature, resulting in an emerging typology of governance relationships underpinning intermodal transport and logistics. Rather than basing public support for intermodal transport on cost reductions from ideal scenarios of full regular trains, this typology aids identification of the internal and external operational models that determine the success or otherwise of intermodal transport services. This framework can be used by future researchers to disaggregate different governance relationships as a precursor to analysing the likely success of new developments.

Institutional Analysis of Governance at Intermodal Terminals and Logistics Platforms

Introduction

This section will draw together the issues raised in the three empirical chapters, in the context of other literature, in order to explicate the key institutional issues relating to intermodal transport and logistics. While the three empirical chapters each addressed a separate element (terminals, logistics and corridors), this section will be structured by the institutional issues, thus treating intermodal transport and logistics as a unified topic, addressed by the four main components derived from the literature in Chapter 3.

Chapter 4 showed that ports can actively develop inland terminals, and differences exist between those developed by port authorities and those developed by port terminal operators. Furthermore, differences can be observed between those developed by ports and those developed by inland actors. Chapter 5 revealed that while rail remains a marginal business, while the industry remains fragmented, while third-party consolidation is not pursued and while fragile government subsidy is still the basis of many flows, intermodal corridors cannot become instruments of hinterland capture and control for ports. Chapter 6 found that in many cases institutional design will constrain integration between maritime and inland transport systems. The conflict between legitimacy and agency creates barriers and if an infrastructure for collective action is not in place (and it is usually predominately a public infrastructure for collective action), then private firms will not act, thus challenging attempts at port regionalisation and keeping the maritime and inland spaces separate. The multi-scalar formal and informal planning regimes in which each port is situated mean that generic port development strategies based on assumptions of hinterland integration will face several regionally-specific challenges. The different operational models identified in these case studies will now be explored in more detail through a classification and analysis of governance relationships.

This chapter will identify the governance relationships existing between intermodal terminals, logistics operations and intermodal corridors. The analysis begins with the development and ownership of intermodal terminals and logistics platforms and expands into logistics practice and port-hinterland corridor integration. The four research topics to be addressed were derived in Chapter 3:

1. Roles of the public and private sectors in processes of planning and development;
2. The relation between the original developer and the eventual operator, including selling and leasing;
3. The relationship between the transport and logistics functions, and other issues internal to the site;
4. The site functions and operational model, including the relationships with clients and external stakeholders.

Each will be addressed in its own section, in which cases from this book and many from the literature will be considered, leading to a final typology that will classify the main governance relationships.

Planning and Development

The main issues identified in the case analysis in Chapter 4 were the role of government planning and funding, whether the developer is from the public or private sector and the eventual role of the site developer in transport and logistics operations. It was shown in Chapter 3 that governance has rarely been addressed directly in the intermodal literature; this section will reveal that it has been raised indirectly through discussions of the role of government in supporting developments, the role of real estate developers and the use of public-private partnerships with varying levels of public involvement. It was also established in Chapter 3 that governance is concerned with managing resources and relationships in order to achieve a desired outcome; this section will proceed by identifying the key stakeholders and the relationships between them.

The efficacy of public investment in terminals has been questioned, considering the difficulties of economically viable operation once the site is built

(Höltgen, 1996; Gouvernal et al., 2005; Proost et al., 2011; Liedtke and Carrillo Murillo, 2012). Yet the example of Verona in Italy in Chapter 4 (owned jointly by the town, province and chamber of commerce) showed that government bodies can be the direct developers of inland freight facilities. An example from the literature is Falköping in Sweden developed by the municipality (see Bergqvist, 2008; Bergqvist et al., 2010; Wilmsmeier et al., 2011; Monios and Wilmsmeier, 2012a). This fully-public model is unusual and depends on the competencies of the public bodies in question. The risk is whether the site can then be leased or sold on to a private operator. Government involvement is more commonly achieved either as a PPP (e.g. Bologna, Italy – Chapter 4), a one-off funding grant or land provision (e.g. Coslada, Spain – Chapter 4; Uiwang, Korea – Hanaoka and Regmi, 2011; Jinhua, China – Monios and Wang, 2013), or through a concession not simply to operate a site but to build it as well (Tsamboulas and Kapros, 2003), through, for example, BOT, DBOT, DBOM or BOOT models (e.g. Nola, Italy – Chapter 4; Lat Krabang, Thailand – Hanaoka and Regmi, 2011).

Developments driven by the public sector run the risk of over-supply; this problem is less prevalent in North America where the private sector focus on profit tends to act as a method of regulating such optimism bias (Notteboom and Rodrigue, 2009a; Rodrigue at el., 2010). Alternatively, public sector developments are more likely to adhere to planning strategies such as location in brownfield sites or economically undeveloped areas. In principle, private sector developments are subject to the same planning approvals, but in practice they often succeed in evading such restrictions (Hesse, 2004), partly due to a lack of institutional capacity to manage planning conflicts (Flämig and Hesse, 2011). Even when local planning rules are applied, the lack of a coordinated regional approach can lead to several undesirable consequences; these include suburban sprawl of logistics platforms (Bowen, 2008; Dablanc and Ross, 2012), a lack of incentive to invest (Ng et al., 2013) and a split of scale economies across institutional jurisdictions (Notteboom and Rodrigue, 2009a; Wilmsmeier et al., 2011).

Logistics platforms are generally more attractive economically than intermodal terminals, therefore real estate developers are more likely to pursue the former. This is especially the case in countries where the public sector traditionally has less direct involvement, such as the USA and the UK. For example, global company ProLogis, in conjunction with CenterPoint, developed the BNSF Logistics Park in Chicago, within the boundary of which was situated a large intermodal terminal developed by rail operator BNSF (Rodrigue et al., 2010). As the freight industry moves increasingly into the private sector, even in continental Europe, this private sector model is becoming more common; for instance the Daventry UK logistics platform and intermodal terminal built by ProLogis (Chapter 5), or the Magna Park development in Germany studied by Hesse (2004).

The analysis of the logistics real estate market by Hesse (2004) showed how the old model of high ownership levels, primarily local firms, few speculative developments, 10 year leases and a weak investment market has changed. The current market shows an increasing share of rental sites, international developers, speculative development, shorter leases of 3–5 years and a strong investment market for new developments. Other authors have demonstrated that average warehouse size has increased in both the UK and the USA (McKinnon, 2009; Cidell, 2010), as is the tendency to agglomeration, with companies locating their DCs within large logistics parks (McKinnon, 2009).

Real estate and public sector developments may be classified together as sites that are intended to be sold or leased to operators. Other sites are developed directly by the eventual operator for their own use. Most rail networks in Europe were managed by the national government until recent times (Martí-Henneberg, 2013); terminals were, therefore, developed both by private transport operators attached to the national network and by the national rail operators themselves. These sites are now mostly owned and/or operated by private operators, as demonstrated by the discussion of the British case in Chapter 5 and the liberalised EU environment on the continent in Chapter 4 (e.g. the vertically-separated and quasi-private but still nationally-owned rail operator managing the Muizen terminal in Belgium). In other countries, the rail operations remain wholly or predominantly under state control (e.g. many terminals developed by state-owned Concor in India, even though private sector involvement is now allowed – Ng and Gujar, 2009a, b; Gangwar et al., 2012). Rail in the United States is privately owned and operated on a model of vertical integration; intermodal terminals are, therefore, developed and operated by the private rail companies (e.g. Joliet intermodal terminal Chicago built by BNSF – see Rodrigue et al., 2010). Intermodal terminals can also be developed by port actors, whether port authorities (e.g. Coslada, Spain – Chapter 4; Enfield, Sydney – Roso, 2008) or port terminal operators (e.g. Venlo, Netherlands – Chapter 4, Hidalgo, Mexico – Rodrigue and Wilmsmeier, 2013).

Ownership and Operation

Whether the site, once developed, is then leased on the landlord model or sold to an operator is the next governance aspect to consider. When sites are developed by a real estate operator the aim is to earn profit through selling or leasing either the entire site or individual plots. Sites developed by government have a contrasting motive; in these cases the decision of selling or leasing is tied to obtaining social benefits. A secondary element of this decision is whether the site as a whole is being disposed of or only individual plots. A real estate developer is likely to lease or sell individual plots, whereas a public body is more likely to lease the entire site to an operator who will then manage the plots.

As shown in the various PPP models examined in Chapter 4, the government role in operations is dependent to some degree on whether or not the government (or other public sector) body has direct involvement in the site or just a shareholding. In the fully-public example of Verona (Chapter 4) the site is managed by an arm's length company established by the public shareholders, so they are not directly involved in day-to-day running. In other cases, the public owner actually operates the site, at least on a supervisory 'tool port' basis (e.g. Shijazhuang, China –

Beresford et al., 2012), although such examples are rare. More likely is that the public body fully owns the site but tenders the operation to a private operator on the landlord model (e.g. Coslada, Spain – Chapter 4; Birgunj, Nepal – Hanaoka and Regmi, 2011).

The decision by a public sector developer of whether to lease or sell is derived to a large degree from the initial motivation for developing the site. For instance, what problems with the previous system are the public sector stakeholders trying to solve by investing in, owning or operating an inland freight node? It is likely to relate either to economic development from supporting the logistics sector, thus bringing jobs and economic activity to the region, or seeking modal shift from road to rail to produce a reduction in negative externalities such as congestion or emissions. The case analyses in this book demonstrate that it is not possible to guarantee such outcomes simply by building an intermodal terminal or logistics platform. In order for the site to be successful in developing intermodal traffic, several operational and institutional barriers need to be overcome. That is why a governance analysis, if it is to be of use to public sector planners and funders, must go beyond the simple issue of ownership; the relation between transport requirements and logistics operations is an essential part of terminal success, and should be understood as part of a decision to approve or fund an intermodal terminal. These issues will be considered in the following section.

Internal Operational Model

In some cases, both the intermodal terminal and the logistics platform may be operated by a single operator. This could, in principle, produce synergies between the two, but in practice this dual role may not align with the core competency of that single operator. For example, a rail operator operating a joint terminal and logistics platform would be different from a 3PL operating that same joint site. The reality is that, even where there is a nominally unified organisational structure encompassing both logistics platform and intermodal terminal (e.g. the Italian freight villages in Chapter 4; Xi'an, China – Beresford et al., 2012; BNSF Logistics Park Chicago – Rodrigue et al., 2010), the intermodal terminal and individual parts of the logistics platform will be operated by different organisations. Each of these organisations may already be operating in the market or they may be established purely to fulfil that role, often with part-investment from the overall site owner. This business model was demonstrated in the Italian freight villages in Chapter 4, where the operator of the rail terminal within the freight village was often a newly-established company with part-investment from the freight village operating company and a rail operator. As the development process for such large projects is capital intensive and entails some risk regarding its eventual operational success, a joint development process is likely, including a real estate developer, rail operator and a public authority. The resulting site, however, will tend to be managed separately by the rail operator (terminal) and real estate developer (logistics platform) (Rodrigue and Notteboom, 2012).

Once the site is operational, the intermodal terminal and logistics platform will commonly be operated separately but may still retain close operational relations. The variety and informality of such relations is difficult to capture in a governance analysis. Venlo, Netherlands (Chapter 4) is a good example of close relations between terminal and logistics, with the terminal operator holding a 50 per cent stake in the logistics platform. Of the five Italian freight villages examined in Chapter 4, in all cases the intermodal terminal was operated by a separate operator from the logistics platform; however, in most cases the logistics platform operator had a high proportion of investment in that rail terminal operating company. In most cases, as discussed earlier, the terminal operating company had been set up specifically to operate that terminal, with ownership from the logistics platform and a rail operator. These examples demonstrate the 'organisational implant' concept discussed in Chapter 3 (Grawe et al., 2012), which increases synergies by placing a representative of one organisation within the other. Further operational integration is possible in the container shunting operations between the terminal and the individual warehouses and distribution centres (both within the logistics platform and in the surrounding area). Normally, this would be arranged by the shipper or freight forwarder but it could be managed directly by the logistics platform operating company through a dedicated shunting operation to serve site tenants and other nearby locations, thus increasing operational integration and lowering costs.

Sites developed without direct involvement of an operator face the danger of optimism bias (Bergqvist et al., 2010). In order to understand the business models of sites that are developed by operators, the next section will compare the different operational models observed in this research with cases from the literature to reveal the governance relationships underpinning these models.

External Operational Model

Rodrigue et al. (2010) classify the primary functions of intermodal terminals into satellite terminal, transloading site and load centre.

A satellite terminal is usually located close to a port (see also the short-range dry port model of Roso et al., 2009) and used to overcome congestion by moving containers quickly out of the port area for processing at the inland location (Slack, 1999; Roso, 2008). There is, therefore, generally a high level of operational integration. The operational focus of the close-range site is often to fulfil administrative tasks, including but not limited to customs clearance; this means that the valuable and congested port land is reserved for container handling functions rather than becoming congested with trucks and containers waiting to complete administrative tasks. Examples of satellite terminals in the literature include Enfield, Sydney (Roso, 2008) and Beijing, China (Monios and Wang, 2013). From a transport perspective, the short distance between the port and the satellite terminal means that the mode used is more likely to be road; rail or barge, however, are also used (e.g. the so-called 'container transferium' recently developed at Alblasserdam just outside the port of Rotterdam – van Schuylenburg and Borsodi, 2010). A road-linked terminal would seem to ignore the main function which is to overcome congestion, but such a model can reduce congestion by reducing the time each truck spends in the port on administrative matters. Moreover, if the trucks are run by a dedicated shunting service controlled by the port terminal operator, this level of integration keeps the move internally controlled and requires fewer gate formalities.

A transloading centre is a site dedicated to changing mode or interchanging between services within the same mode. An example is the IFB Mainhub terminal at the port of Antwerp, Belgium, which reconfigures wagon sets for trains interchanging between several port terminal yards and several inland origins and destinations (Macharis and Pekin, 2009). A transloading site could, in principle, be just the terminal with no services or storage nearby, but in practice it generally involves such services; pure transloading sites are rare. Thus, while the primary function as per the definition is interchange rather than servicing a local market, in most cases it normally does so in order to make the site economically feasible, which leads into the third main function, that of a load centre.

The load centre concept, as applied to transport terminals, refers to a large intermodal terminal servicing a large region of production or consumption. It is, therefore, the classic kind of inland node as it serves as a gateway to a large region and is more likely to be set within a specific logistics platform or in an area with high demand for such services. The load centre concept fits well within the American inland port typology, generally referencing a large site with a logistics platform located either nearby or as part of an integrated site. Examples include Rickenbacker Inland Port (as discussed in Chapter 6) and BNSF Chicago (Rodrigue et al., 2010). However, the freight village case studies in Chapter 4 revealed that, even within integrated warehouse and terminal sites, the majority of usage for a terminal does not often come from tenants of the freight village. Similarly, McCalla et al. (2001) showed that it is often merely a minority of businesses in a region that utilise the intermodal facilities located there. Agglomeration properties of a load centre can therefore be said to be rather weak.

A fourth operational model must be included, as it relates all three of the above functions, in particular the satellite terminal and load centre. The extended gate concept is a specific kind of intermodal service whereby the port and the inland node are operated by the same operator, managing container flows within a closed system, thus achieving greater efficiency. As seen in Chapter 4, at Venlo, the Netherlands, the intermodal terminal is set within a logistics platform and the operator of both the port terminal and the intermodal terminal also holds a 50 per cent stake in the logistics platform (for more detail see Rodrigue and Notteboom, 2009; Rodrigue et al., 2010; Veenstra et al., 2012; Monios and Wilmsmeier, 2012a). Successful development of an extended gate service requires an operator to overcome several institutional barriers but offers significant opportunity to improve the efficiencies of service planning and, therefore, improve the economic viability of intermodal port shuttles. It also expands the port vs inland distinction highlighted throughout this book, widening into an appreciation of international vs domestic traffic; these tend to have different equipment requirements (see Chapter 5) and can involve conflicts between operational models and priorities between port and inland actors. These issues were discussed in the directional model applied in Monios and Wilmsmeier (2012a) and Monios and Wang (2013).

Logistics platforms, like intermodal terminals, exist in different sizes, offering different services, and can be more or less developed (as discussed in Chapter 2). Some are small, catering to local shippers and offering few services, while others are large sites offering comprehensive value-added services with large volumes and wide geographical coverage. Distribution centres are now integrated elements of the transport flow (see Chapter 2) and their relation with the transport function must be analysed in more detail.

Different operational models can be identified across a range of logistics platforms. The analysis in Chapter 4 revealed different models applied in Italian freight villages. In the majority of instances, the site operator (a body set up and controlled by the owners) merely sells or leases individual plots to customers (e.g. Bologna). The tenants may be either individual shippers doing their logistics in-house or, in some cases, a site may have a majority of 3PLs as tenants (e.g. Marcianise). At Rivalta Scrivia, however, the case study revealed how the operator of the logistics platform performs logistics operations directly for the tenants. This more integrated model supports consolidation and feeds the intermodal terminal, with a result that at this site the proportion of traffic at the intermodal terminal belonging to tenants of the logistics platform is far higher than at other freight villages.

The case studies analysed in the empirical chapters of this book have demonstrated the importance of understanding the relationships between the port, the terminal, and the external stakeholders along the intermodal corridor. How does the intermodal terminal relate with rail operators and logistics providers organising company trains? How does the terminal operator interact with port authorities, port terminal operators or shipping lines in managing port shuttles? As discussed in Chapter 3, intermodal corridor operations can be managed in different ways to lower transaction costs, such as contracts, joint ventures and integration through mergers and alliances. Above all, terminal volume is linked to traffic flows therefore the terminal operator requires a close relationship if not some level of integration with the rail operator to guarantee usage (Bergqvist et al., 2010).

The empirical cases in this book as well as those from the literature exemplify different levels of collaboration and integration in intermodal corridors. The intermodal terminal operator may be independent from rail service operation (e.g. Azuqueca, Spain – Chapter 4), it may run rail services for any users (e.g. Delcatrans, Belgium – Chapter 4; Freightliner, UK Monios and Wilmsmeier, 2012b) or it may run rail services directly for the site tenants (Venlo, Netherlands – Chapter 4; Minto, Sydney – Roso, 2008). Similarly, the operator of the logistics platform may do the logistics for site tenants (Rivalta Scrivia, Italy – Chapter 4) or it may not. From a port perspective, there may be investment from a port authority (e.g. Coslada, Spain – Chapter 4; Enfield, Sydney – Roso, 2008) or port terminal operator (e.g. Venlo, Netherlands – Chapter 4; Hidalgo, Mexico – Rodrigue and Wilmsmeier, 2013). Additionally, the relation between the port and the inland terminal may be a highly integrated extended gate style of operation (Venlo, Netherlands – Chapter 4) or it may not (being the majority). Similarly, port actors can be directly involved in establishing intermodal services or corridors (e.g. Venlo, Netherlands – chapter 4; Barcelona, Spain – Van den Berg et al., 2012; Alameda Corridor – Jacobs, 2007; Rodrigue and Notteboom, 2009; Monios and Lambert, 2013b; Eurogate – Notteboom and Rodrigue, 2009a). It will mostly be large ports with the necessary resources that are likely to engage in such tactics, meaning that the levels of integration required for such aggressive hinterland control will be the exception rather than the norm.

Classifying Governance Relationships Between Intermodal Terminals and Logistics Platforms

The preceding discussion, split into four sections, has drawn out the key governance relationships in intermodal transport and logistics, based on findings from the empirical cases in this book, combined with additional cases from the literature.

A site may be developed by a variety of actors: for example, government, real estate developers, rail operators, 3PLs, port authorities, port terminal operators, shipping lines, independent operators and others. Each actor will have different motivations; for example, obtaining social and economic benefits (government), to sell the site or parts within it for profit (real estate developer), as part of an existing business (e.g. rail operator or 3PL) or for hinterland capture (e.g. port actors).

Different relationships were identified between the owner and operator, drawing on previous uses of governance in the transport literature. This part of the typology covers whether the owner operates the site directly, at arm's length, through contracts or via a concession (landlord) or lease.

The third issue discussed above is the main function(s) of the site, in particular whether the site is an intermodal terminal or logistics platform or both. Part of this question includes the nature of the operator of each site (intermodal terminal and logistics platform) and the relation between them. This follows on from the development process (related to the original aim) and points the way towards more specific questions about the operational model.

The fourth issue concerns operations. A number of detailed operational issues have been raised in the literature regarding the performance and economic viability of intermodal transport. The literature has pointed towards varying models of collaboration and integration as potential aids in this endeavour, so this section will classify the different models as part of the governance typology.

The governance relationships identified above are classified in Table 7.1 below.

Table 7.1 Classifying governance relationships between intermodal terminals and logistics platforms

#	Classification	Grouping / Relation	Option 1	Option 2	Option 3	Option 4	Option 5	Option 6	Option 7
1	Developer of each part (terminal or logistics or both)	Built to sell/lease on in order to get benefits (economic and social for govt, profit for RE developer)	Govt body (which level and type)	Real estate developer					
		Built to operate themselves as part of a business strategy (although may have government investment or planning support)	Rail operator	3PL	Port Authority	Port terminal operator	Shipping line	Independent	
		Other	PPP	Other					
2	Operator of each part (terminal or logistics or both)	Relation between owner and operator	Operated directly by owner	Tool (operated directly but with some sub-contracting)	Operated through an arm's length company	Landlord (publicly owned but operated by private company under concession)	Leased	Other	
		Type of operator of each part	Rail operator	3PL	Port Authority	Port terminal operator	Shipping line	Independent	Other
3	Internal operation model (relation between terminal and logistics)	Relation between operator of terminal and logistics	Single operator	Mixed	Separate operators				
4	External operation model (relation with clients and others)	Relation between logistics platform and site tenants	Tenants do own or in-house logistics	Tenants are 3PLs doing logistics for various clients	Site operator does logistics for site tenants				
		Relation between terminal and rail service providers	Terminal operator is independent of rail service operators	Terminal operator runs rail services for all	Terminal operator runs rail services for site tenants				
		Relation between either terminal or logistics platform and ports	No relation with port	Port actor has invested partly or fully in site	Port actor involved in joint service operation				

Discussion

The literature analysis in Chapter 3 explained how port governance studies have focused primarily on the relation between the owner (often a local or regional government) and the operator (commonly appointed by a tendered lease). By contrast, the case analysis in this book has revealed that the key governance aspect of intermodal terminals and logistics platforms is related to operational characteristics. The operational types derive from the relation between the operator and the external actors (ports and rail operators), as well as the relation of the operator to the tenants, and, as reflected in the classification, the relation between the intermodal terminal and the logistics platform.

The importance of these relationships in successful intermodal transport and logistics means that the governance of ownership and operation (covering the landlord issue and tendering) is more important for ports than for inland sites, and forms only the first half of the classification. For inland intermodal terminals or logistics platforms, the ownership matters principally when considering the desired outcome from the investment (social and economic benefits for the public sector actor or profit from selling the plots for a real estate developer). A decision by an industry actor (rail operator, 3PL, port authority, port terminal, etc.) to develop a new site is based not only on short-term profit but a long-term strategic decision to operate the site as part of their larger business.

The classification developed in this chapter extends the ownership aspect to include the operational model, both internal (relation between the terminal and the logistics platform and who operates each) and external (relations with tenants, rail services and ports). These issues are crucial because, just as a port's success (both for itself and for its region) is related to the ability of the owner to negotiate a successful concessionaire that will attract shipping lines, the success of an intermodal terminal is related to many operational aspects such as establishing regular intermodal services, consolidating flows to fill them, and reaching out to decision makers in the maritime sector (ports and shipping lines) to embed the terminal in global flows.

The above approach to governance contrasts with previous work on port governance, because, in the port's case, external relations are less relevant to the original lease decision. It is up to the port terminal operator to run the site in the most profitable way. Moreover, strategies have merged over recent decades so most terminals have links with shipping lines and with other terminals in the same global company. Intermodal terminals and logistics platforms are far smaller concerns than ports; they are less likely to be part of a global or even national portfolio, which means that they exhibit a variety of relationships with rail operators, site users and ports. The three kinds of external relations identified in this research are depicted in the fourth section of the classification in Table 7.1; they can be used as a basis for future research because they are directly related to the ability to achieve successful intermodal services.

The use of the classification is to identify the relevant resources and relationships relating to each model, which can then inform the policy background of supporting intermodal transport services. The requirement for greater internal and external integration as established in this chapter goes some way towards understanding why intermodal terminals do not always achieve the modal shift aims of government policy, despite often large amounts of public investment. There is a danger that governments invest in infrastructure without addressing the governance issue sufficiently, in particular the relation between internal and external integration models. As shown in Chapter 3, the supply chain literature focuses on these issues as part of the business decision to integrate, a practice that is increasingly necessary for firms to gain competitive advantage. It has been recognised for some time in the supply chain literature that competition is increasingly between entire supply chains rather than individual firms (Christopher, 1992), but this realisation is sometimes lacking in the transport literature. It is especially absent from government decisions to fund intermodal terminals and logistics platforms based purely on an ideal scenario promising a reduction in transport costs, in isolation from an analysis of their operational models.

The analysis of cases in this book as well as other cases from the literature has shown that operational models are directly related to whether an operator can succeed in developing intermodal services, based on the ability to cooperate, integrate, consolidate and plan. These are not just operational concerns; in many instances they are derived directly from the governance model. Future research can use this classification to aid identification of the different ways such integration can be pursued, and, more importantly, use this model to recognise a lack of such integration.

Conclusion

As discussed in the introduction, the operational difficulties preventing the economic feasibility of intermodal transport are well known. While some discussions of the role of integration and collaboration have been raised in the literature, the relation between the owner, operator and operational models of freight nodes has been insufficiently addressed, suggesting the need for an institutional analysis of these models. The lessons from the governance literature (as synthesised in Chapter 3) have been applied to a study of intermodal terminals and logistics platforms (expanded from the case studies in Chapters 4–6) in order to classify the role of internal and external operational models in the success or otherwise of intermodal transport and logistics.

The conclusion from this chapter is to recognise the necessity of understanding operational models that provide greater synergies, not only between the users of the intermodal terminal and the logistics platform, but in the relations between the two sites as well as relations with external stakeholders, such as transport providers and port actors. Previous literature focused on classifying sites by ownership (e.g. the

World Bank and port governance models); this is a first step, but insufficient on its own. Those models have been extended in this chapter by adding another two layers derived from the supply chain literature; namely, internal and external integration. This expansion is essential to the understanding and especially to the success of intermodal transport and logistics. If policy goals of modal shift are to be achieved, intermodal transport can no longer be considered in isolation from the logistics strategies in which it is embedded. The clear conclusion is that government funding used for intermodal infrastructure and operational subsidies must be aligned with an understanding of how intermodal flows are embedded within internal and external relationships.

Chapter 8

Institutional Adaptation and the Future of Intermodal Transport and Logistics

Introduction

The final chapter situates intermodal transport and logistics within the wider context of port geography. It begins by drawing on the literature to provide a brief analysis of how the European and North American case studies in this book compare to the geographies of intermodal freight transport across continents. This comparison demonstrates how the spatial and institutional characteristics of intermodal transport and logistics derive from the transport and logistics environments of each continent. The developing economies of Asia, Africa and Latin America are each learning the lessons of previous developments in Europe and North America, but the institutional settings direct how these issues play out.

The chapter then returns these findings to the theoretical context of port regionalisation and hinterland integration. It identifies and discusses challenges to the port's ability to capture or control hinterlands through the strategies of integration that might be expected in an ideal port regionalisation scenario. It is not easy to maintain the favourable operational and institutional conditions required for successful port regionalisation, meaning that port actors need to adapt if they are to manage regionalisation processes successfully. This chapter discusses some recent research on the institutional adaptations of port actors as a result of these challenges, and identifies directions for future research.

Comparing the Geographies of Intermodal Freight Transport and Logistics Across Continents

Academic literature over the past decade has begun to develop conceptual models to classify and analyse different strategies of inland terminal development within the context of port development and hinterland integration. The approaches have proceeded from spatial (nodes, corridors, clustering, load centres and sprawl) to institutional (ownership, investment, stakeholder management and collective action within transport chains). As has been seen throughout this book, the geography of freight transport and logistics operates at the intersection between spatial and institutional analyses.

The dominant focus has been on Europe and the United States (e.g. Rodrigue and Notteboom, 2009; Roso et al., 2009; Bergqvist et al., 2010; Rodrigue et al., 2010;

Monios and Wilmsmeier, 2012a). While in recent years some literature on Asia (e.g. Ng and Gujar, 2009a, b; Ng and Tongzon, 2010; Hanaoka and Regmi, 2011; Beresford et al., 2012; Gangwar et al., 2012; Ng and Cetin, 2012; Lu and Chang, 2013; Monios and Wang, 2013), Africa (e.g. Garnwa et al., 2009; Kunaka, 2013) and Latin America (e.g. Padilha and Ng, 2012; Ng et al., 2013, Rodrigue and Wilmsmeier, 2013) has begun to be published, it remains the case that a geographical understanding of the spatial development of intermodal freight transport in developing economies has been insufficiently established (Ng and Cetin, 2012; Notteboom and Rodrigue, 2009a).

The hinterland freight geography of North America represents a landbridge and Europe is based on coastal gateways and inland load centres (Rodrigue and Notteboom, 2010), while the East Asian hinterland model has been categorised as coastal concentration with low inland coverage (Lee et al., 2008). European and North American seaports are generally conceptualised as increasingly integrated with their hinterlands, as per the regionalisation model discussed throughout this book, but the historical lack of inland penetration of Asian and Latin American ports would suggest that such hinterland integration models do not apply there. While this appears to be true in India (Ng and Cetin, 2012) and Latin America (Ng et al., 2013), in China this spatial pattern is being altered by, for example, the establishment of several inland terminals in the Chinese hinterland over the last decade.

Monios and Wang (2013) found that the inland terminal network emerging in China reflects similarities to patterns observed in more integrated networks such as Europe and North America. Over the last decade, major Chinese seaports have paid increasing attention to the need to develop inland terminals due to the competitive pressures of overlapping hinterlands with neighbouring ports. A surge in investment in port infrastructure over the last ten years, which was primarily intended to facilitate the expansion and improvement of cargo handling capacity, has resulted in excessive inter-port competition. Cullinane and Wang (2012) argued that this investment may not be sustainable and could lead to an inefficient utilisation of port resources if the market environment were to change fundamentally. It is especially true for China as its export-oriented economy faces the simultaneous threats of a rapid rise in domestic labour costs and a contraction in global demand. It is therefore essential for port authorities and terminal operators to maintain their growth by securing traffic flows, balancing the dependency on exports and/or enhancing their hinterland supply. The latter is an easier choice, which goes some way towards explaining the observed strategy of investing in inland terminals.

Inland cities in China have also evinced significant interest in this concept with the purpose of enhancing the competitiveness of their local economies. However, as seen in some of the cases studies in this book, the strategic (to compete with other ports) and operational (to improve access to the port for users) aims of the seaport do not always align with the policy aspirations of central government or the planning strategies of local governments. These conflicts materialise in operational problems such as the inability to match wagon and container configurations for maritime flows (the primary interest of port actors) with those required for domestic flows. This situation can be compared with previous experience in the USA, where the majority of container movements are domestic, as may be the case one day in China.

Similarly, recent research indicates that the hinterland strategies of globalised port terminal operators in Central America exhibit some replication of European port-led strategies also (Rodrigue and Wilmsmeier, 2013). One of the most successful cases of Outside-In development discussed in this book is the extended gate inland terminal development at Venlo in the Netherlands (see also Veenstra et al., 2012). This site was developed by ECT (a subsidiary of HPH), a private port terminal operator situated within the port of Rotterdam. Rodrigue and Wilmsmeier (2013) examined the case of the port of Vera Cruz in Mexico, where port terminal operator HPH is pursuing a similar case of Outside-In development. As is often the case in developing countries, this case experienced difficulties making rail transport feasible and security concerns prevent a successful establishment of inland customs clearance. It is therefore primarily institutional issues that are constraining the attempts to establish successful intermodal transport and logistics.

Different models of inland terminal development relate partly to the motivations for using the site; these motivations are not always the same in developing and developed countries. One motivation is to reduce transport costs by bundling flows to achieve economies of scale on key routes, whereas another aim is to reduce transaction costs by moving administrative activities such as customs inland. The inland terminal may be far inland or it may be close to the port, thus serving purely to reduce costs associated with congestion-related delays, both for the port operator and the shipper.

As seen in this book, related logistical services are also important to successful inland terminal operations, especially so for developing countries. Containerisation can be a problem in developing countries where trade is based more on raw materials and bulk movements therefore containerised trade and all the attendant services are less developed. Shippers may have to drive a load to the port and wait a significant amount of time to get it containerised before it can be loaded on the ship. All of these costs reduce the competitiveness of exports from such countries. Meanwhile, containerisation of bulk cargos (e.g. grain) has emerged as a new tendency in enlarging the container transport market (Rodrigue and Notteboom, 2011) and it has been utilised in the north east of China for services to the port of Dalian. Container management can also be a problem for developed countries with long distances. For instance, in the USA shipping lines are reluctant to send maritime containers far inland as they cannot guarantee an export load for the return journey (as the USA is an import economy) (Monios and Lambert, 2013b). So they tend to transload at the port from 20ft/40ft deepsea containers into 53ft domestic boxes (Notteboom and Rodrigue, 2009b).

Inland customs clearance is a more significant issue in developing economies than for European or North American inland terminals. Inland Clearance Depots

(ICDs) appeared in Europe from the 1960s (Garnwa et al., 2009), as the container revolution and motorway development changed the transport geography of freight distribution from the coast to inland centralised locations. Specifying an inland location on the bill of lading and clearing customs inland were therefore attractive to shippers. Now, with the ease of electronic documentation, shippers can clear customs at a point of their choosing with less impact on the practicalities of their business. The Authorised Economic Operator (AEO) system in Europe allows operators thus designated to proceed without the need for physical inspection. In addition, the union of customs across the European Union in addition to a currency union across much of Europe have simplified the process enormously. In the United States, as a single country, customs jurisdiction and currency are unified; moreover, as 89 per cent of freight is domestic (FHA, 2010), international freight movements are less integrated with purely domestic flows and thus have less incentive to clear customs inland.

In both Europe and the United States, joint customs jurisdictions and a single currency make customs clearance procedures simpler, faster and cheaper, as fewer organisations are involved and currency conversion expenses are eliminated. In countries where this is not the case, being able to perform administrative duties (including but not limited to customs) inland can produce significant cost savings.

In Africa and China there are more small shippers who can gain from streamlined customs and administrative procedures. This is particularly the case for landlocked African countries where additional barriers exist between the port and the hinterland location (Adzigbey et al., 2007). Much work has been done at the international level to harmonise customs procedures and obtain some of the benefits of customs unions such as the EU. Being able to clear customs later (for imports) can improve cash flow by paying import duties when the goods arrive inland (or more specifically, when they are taken from the inland terminal, giving the potential for extended storage, because the fee is paid only when the goods are required). The downside of this system is that funds may be tied up in customs bonds, but attempts are being made to establish an authorised importer system whereby reporting requirements are reduced and a bond may not be requested (Arvis et al., 2011). Europe and the United States have more mature transport and logistics sectors, thus firms tend to be much larger, with higher cash flow.

Since 2006, shippers in China have benefited from customs reform, through the establishment of an A or AA certificate, which is a credit rating by customs authorities for those companies that have (a) been registered more than two years, (b) achieved at least US$500,000 import and export value (US$1,000,000 required before 1st April 2008), and (c) a history of observing customs laws, rules and regulations. Eligible companies can clear customs inland at an inland terminal and their goods are transported between the port and inland location by a supervised road fleet or block train.

In terms of the cost benefits of intermodal transport, the physical geography of a country like China means that its potential to develop intermodal transport has similar advantages to North America, where rail is competitive with road due to long distances and the ability to run long trains with double stack capacity. In particular, the vertical integration of rail in the United States means that transaction costs are lowered and investment in rail infrastructure is more directly related to service development and operational requirements. Moreover, as the economies of scale are greater in the United States, rail operators focus on their core business of transportation, running large sites with rail throughput exceeding 100,000 TEU annually (Monios and Lambert, 2013b). Containers are then taken elsewhere for logistics activities, and 'co-location' of transport and logistics activities is less common (Rodrigue and Notteboom, 2010).

In Africa, road remains dominant because even inland locations that have a rail connection have struggled to attract traffic due to a number of reasons (Nathan Associates, 2011; Kunaka, 2013). Despite long distances, inefficient rail operations and poorly maintained infrastructure mean that, first, the rail costs are much higher than they would be elsewhere and second, even when they are lower, the inconvenience and unreliability outweigh the savings in transport costs, as more time has to be built into the supply chain and higher inventory levels are required for stock buffering (Arvis et al., 2007). Similar fragmentation and lack of service provision are preventing rail competing with road in China at both short and medium distances.

Inland terminal development in Europe is often subsidised on the basis of benefits to be gained from modal shift of freight from road to rail, whether societal benefits from reducing pollution and congestion or benefits for industry such as reducing transport costs for shippers (Wilmsmeier et al., 2011). However, the economic feasibility of intermodal transport remains challenged due to short distances, inability to double stack and the inevitable fragmentation resulting from the multiple institutional jurisdictions covered by the European rail network. In order to attract enough custom to fill trains, many operators offer a door-to-door service, often rebranding themselves as logistics providers and taking more direct involvement in container management through different strategies (Monios and Wilmsmeier, 2012a). However, fierce port competition has resulted in overlapping of market coverage thus splitting potential scale economies. Similar dangers can be observed in China, as some inland terminals have dedicated agreements with specific ports, while others have multiple agreements; it could therefore be the case that competition will dilute the potential consolidation of traffic. As a consequence, successful intermodal transport may be challenged except on longer routes with double stack access (like in the USA).

Thus the spatial development of intermodal transport results from both transport and logistics requirements, which are often different in developed and developing economies. In particular, the institutional constraints in different continental settings exert a significant impact on successful intermodal transport and logistics. While some evidence is available in developing countries of replication of the hinterland integration strategies already exhibited in developed countries, closer investigation reveals that different aspects such as the pros and cons of rail transport, or the logistics requirements of containerisation, container availability, clearing customs, stock buffering, etc. all influence transport decisions in different ways. In developing countries it is mostly logistics improvements that are required, through trade facilitation measures, business development strategies, customs harmonisation and so on. These issues need to be resolved before intermodal transport can work to its strengths of consolidated, scheduled long haul shipments. Even in Europe, with many administrative hurdles cleared and a currency and customs union, the institutional challenges of intermodal transport remain formidable in any but the most dense high demand corridors. That is why the USA remains the world leader in intermodal transport. China has the potential to replicate this success, but several limitations must first be overcome.

Intermodal Transport and Logistics within the Wider Context of Port Geography

Each of the three empirical chapters highlighted conflicts between maritime and inland actors, a lack of integration, institutional barriers, and the importance of understanding the specificity of market structure and the limitations of political design; each of these limit the extent to which port regionalisation processes can occur. However, the previous discussion of intermodal transport and logistics across different geographical settings demonstrates that these issues remain context-dependent, thus challenging the ability to capture them in a single concept.

Rodrigue et al. (2010: p.2) noted that 'the inland port is only an option for inland freight distribution that is more suitable as long as a set of favourable commercial conditions are maintained'. Similarly, it might be said that port regionalisation requires a set of favourable commercial and institutional conditions to be maintained. The findings from the cases analysed in Chapters 4–6 of this book suggest that it is not easy to maintain such conditions, and the analysis of governance relationships in Chapter 7 illustrated some of the institutional diversity underpinning such conditions.

The trends towards port devolution and the deregulation of transport services discussed in Chapter 2 have increased possibilities for the private and public sectors to cooperate, which, as shown in Chapter 7, can take different forms. Land use and transport planning require integrated approaches across local, regional and national boundaries in order to influence and direct port development, including inland investment strategies. Both the discussion on terminal development in Chapter 4 and the operational discussion in chapter 5 highlighted the difficulty of making intermodal transport feasible, especially in Europe. Many terminals had their development subsidised by the public sector, and many operators still receive public funds. Many rail operators in Europe continue to receive subsidies from their national governments; this subsidy indirectly supports the small terminals that continue to exist.

Sixteen years ago, Höltgen (1996) raised concerns with the proliferation of freight terminals that were not part of a strategic plan, and these issues persist.

More recently, Bergqvist and Wilmsmeier (2009) noted that inland terminals are being developed on an ad hoc basis and this development could threaten their efficiency and hence potential for modal shift. They suggested that government policy could be required to enable a planned system of inland terminals in ideal locations linked by high quality transport infrastructure. The requirement for any facility that benefits from such legislation would be that they remain a common-user facility and publish a transparent pricing structure.

Woxenius and Bärthel (2008) discussed the idea that terminals could be considered infrastructure rather than operations, thus placing them within the government's sphere of influence, which would make it simpler for them to be implemented through government subsidy and then operated by the private sector. Ng and Gujar (2009a, b) discussed different measures taken by the Indian government to direct inland terminal development, some resulting in artificial transport chains that would not otherwise exist. On the other hand, non-intervention has been questioned as well. Rahimi et al. (2008: pp.363–4) noted that 'there is now increasing recognition, from non-planners and the private sector alike, that the "free market" approach to logistics channel formation is not going to work efficiently in the future, especially in large metropolitan areas'.

Local and regional planning authorities, therefore, must improve the integration of their transport planning with industry needs, whether that be market demand or operational requirements. Yet this aim is complicated by the fact that the funding is often provided to local and regional scales by national governments. Chapter 6 highlighted the difficulties of managing such institutional relationships across spaces and scales. In particular, the importance of informal regional cohesion across devolved governance spaces was demonstrated, a finding of particular relevance to the European context, where subsidies can be local, regional, national or supranational (i.e. the European Union).

Port actors do not generally possess the institutional capacity to drive developments far beyond their perimeter. This is more obviously the case for port authorities, which are more likely to be working on a public mandate from the city or region, but it also applies to private port terminal operators. Their institutional design (e.g. a board of directors who report to the shareholders) is generally focused on the core competency of container handling, and this structure is unlikely to be aligned with the relevant business sectors engaged in purchasing land and dealing with the regulatory and other issues of developing a subsidiary in the hinterland. Short-range satellite terminals for overspill functions may be feasible, but the port development strategy of the decision makers in the port terminal operator is not generally compatible with the development of load centres hundreds of miles away. However, the cases in this book and in the literature have shown that some port terminal operators have successfully invested in hinterland terminals. As the port regionalisation concept focuses primarily on the port authority, it devotes insufficient attention to the role of the port terminal operator, which, as recommended by Slack and Wang (2002), must be considered an essential part of any new spatial model of port geography.

Wilmsmeier et al. (2010, 2011) introduced a conceptual approach to inland terminal development, providing a directional focus absent in the priority corridor model of Taaffe et al. (1963) by adapting the terminology of industrial organisation (i.e. forward and backward integration). In this model, the authors contrasted Inside-Out development (land-driven e.g. rail operators or public organisations) with Outside-In development (sea-driven e.g. port authorities, terminal operators). This approach identifies how different institutional frameworks reveal nuances in the different kinds of integration between inland terminals, logistics platforms, rail operators and seaports.

The Spanish cases in Chapter 4 exemplify both kinds of development. Outside-In is followed as a business strategy by port authorities seeking to access an inland market, while the Inside-Out model is pursued by regional authorities seeking to bring development to a region. The priority of securing hinterlands for the ports was joined with the inland priority of providing a high capacity link for local and regional shippers, as well as the economic development aims of the local and regional authorities.

The Venlo case in Chapter 4 demonstrates an Outside-In development, driven by the private port terminal operator. It can be viewed as an example that fits the port regionalisation concept because of its integration of operations. Venlo represents the kind of hinterland access strategy that many ports and inland locations would like to copy, but many institutional, operational and legal difficulties exist (Veenstra et al., 2012). The supply chain literature reviewed in Chapter 3 showed how analyses of integration must consider not merely ownership or investment but process integration, both internal and external. These issues are clear in the Venlo case, because it is not simply the joint ownership but the closed loop run between the port and the inland location that provides opportunities for higher efficiency than is possible under other arrangements. The full visibility of containers through PCS and the direction of such movement by the staff at the inland location provides a clear advantage compared to individual rail operators trying to achieve these efficiencies with separate streams of traffic.

The Italian freight villages examined in Chapter 4 are clear examples of Inside-Out development, as their logistics/warehouse/freight village activities are land-focused. The case study analysis showed that the freight village concept is good for logistics but has had very little success integrating with ports. Indeed, even getting rail traffic at all is not easy for the sites located in the peripheral south due to the road-dominated and fragmented Italian logistics system, and some freight villages have very large intermodal terminals with very low rail traffic. An evocative comment by one interviewee was that 'rail does not exist in Italy'.

The Inside-Out model tends to be publicly driven, as a means to attract investment and development to a region, particularly since a key role of the public sector in infrastructure development is to lever in private sector investment. It is often the case that the profit potential for the private investor in intermodal transport developments is unclear, therefore it can be difficult to attract them. Developing such infrastructure allows container flows to be bundled on high capacity links such that private operators can then bid on this consolidated traffic. By creating an infrastructure corridor that did not exist previously, a barrier to entry is removed (in the form of a large upfront investment) that had kept private operators out of the market. Therefore, the Inside-Out model tends to involve different companies working together in strategies of low integration, whereas Outside-In generally involves greater cooperation since it aims to increase access and/or efficiency in a more mature marketplace.

While Outside-In can be generalised as a privately-driven enterprise to protect or develop business for an existing company, such companies may be publicly owned. So ECT in the Netherlands is a private company integrating inland, while in Spain it is the publicly-owned port authorities which, nonetheless, act as profit-making companies. Looking at the cases in Chapter 6, the Alameda Corridor represents an Outside-In strategy of a port authority integrating inland (although only to the extent of the infrastructure corridor rather than the inland terminals), while the Rickenbacker inland terminal is an Inside-Out strategy, developed by rail operator Norfolk Southern, which, nonetheless, has good relations with the port of Virginia through the larger Heartland Corridor project.

Even with Outside-In development, inland organisations remain heavily involved. While, for example, the port authority or terminal operator may be considered to drive the process and thus the direction, in reality they will be forming partnerships with inland operators or terminals, rail services, logistics providers and other actors. So in the case of Spain, the ports only have a minority shareholding and it is really the inland terminal operators who direct operations, despite the process being driven initially and marketed heavily from the seaward side. Likewise, in the Netherlands, ECT has integrated inland to manage their container flows, but they work in partnership with a logistics company to operate that aspect of the inland site. Thus, even when a port actor may be said to pursue inland integration as a strategy of port regionalisation, ports still retain a primary focus on their core business. This focus on the core business of port and shipping operations is evident in recent trends away from inland integration. The economic challenges of the current recession have led some carriers to divest themselves of their inland transport holdings, in addition to transforming their in-house logistics divisions into standalone profit centres. One notable example was that in June 2013 Maersk sold their European rail subsidiary ERS to Freightliner (Wackett, 2013).

This distinction is a shorthand way of identifying potential strategy conflicts between actors with different motivations. Ports invest in inland terminals to capture and control hinterlands as well as to push containers inland to alleviate port congestion, subject to the ability of the port terminal to act directly in rail operations through joint ventures or similar business models. Rail operators develop terminals and port shuttles for similar reasons, yet need to integrate international and domestic flows with different container and wagon requirements and other planning difficulties. Government agencies, whether local, regional or national, develop terminals generally from business development motivations, therefore often include a logistics platform. The motivation of the port actors (whether port terminals or port authorities) are thus different to inland actors, and require complementary operational models in order to work successfully. Similarly, Hesse (2013: p.40) contrasted the local and regional interests and influence of city stakeholders with the regional, national and international focus of port actors: 'the two different groups of stakeholders involved are quite distinct in their power and in their potentials to achieve their goals'.

Wilmsmeier et al. (2011) argued that this differentiated perspective had not received sufficient attention in discussions of the port regionalisation concept. This model has since been used to aid disaggregation of regionalisation strategies and comparison of potentially conflicting strategies that may be pursued by terminals within a port or between ports within the same range. Ng and Cetin (2012) suggested that Inside-Out development is the common model in developing countries, as opposed to Outside-In in developed countries, whereas Monios and Wilmsmeier (2012a) showed that Inside-Out development is common in developed countries also. Increasing port competition in China has spurred several Outside-In developments there (Monios and Wang, 2013).

Both the cases in this book and those in the literature analysed in Chapter 7 revealed the lack of hinterland integration and the limitations of the port to act, elucidating several potential restrictions on their ability to control or capture hinterlands through the strategies of integration that the port regionalisation concept suggests. While Notteboom and Rodrigue (2005) are correct to some degree to state that regionalisation is 'imposed on ports' (p.302) by landside actors, and that 'the port itself is not the chief motivator for and instigator of regionalisation' (p.306), an inherent contradiction is revealed, in that the concept definition also asserts the importance of port competition and the requirement for port actors to capture and control these emerging inland networks, for example by developing an 'island formation' in 'the natural hinterland of competing ports' (p.303). The limitations of the port actors to do so, as identified above, particularly reflect the fact that it will mostly be large ports with the necessary resources that are likely to engage in such tactics. This means that the levels of integration required for a true regionalisation process will be the exception rather than the norm.

The port regionalisation concept requires adjustment to highlight the opportunities and barriers for successful port regionalisation, and recognise inherent difficulties due to the nature of inland freight operations. It is not easy to maintain the favourable operational and institutional conditions required for successful port regionalisation, meaning that port actors need to adapt if they are to manage regionalisation processes successfully. Some recent developments show that some ports are indeed adapting to this problem and changing their ability to act, so that they can actively engage in such strategies, which will be considered in the following section.

Institutional Adaptations of Port Actors[1]

The inland terminal case studies in Chapter 4 demonstrated that land-driven or Inside-Out terminal developments have generally been the most common as port actors have not possessed the institutional capacity to drive developments deep in the hinterland, but this situation is changing. Port congestion and fierce competition for overlapping hinterlands are forcing port authorities and terminal operators not only to take investments but even to drive such developments themselves (Monios and Wilmsmeier, 2012a; Monios and Wang, 2013). If they are to pursue such Outside-In developments, port actors must expand their institutional capacity beyond the current system focused on a core competency of container handling. Their institutional designs can be transformed through, for example, processes of privatisation or corporatisation (Notteboom and Rodrigue, 2005; Ng and Pallis, 2010; Sanchez and Wilmsmeier, 2010; Jacobs and Notteboom, 2011; Notteboom et al., 2013). From a theoretical perspective, this process has been addressed through relational (Jacobs and Notteboom, 2011), territorial (Debrie et al., 2013) and combined (Monios and Wilmsmeier, 2012b) approaches.

Institutional analysis of port actors explores constraints on their ability to act, stemming from their specific nature. Port development is characterised as path dependent, heavily constrained by past actions and institutional design, but also contingent, in relation to private investment and public planning. Ng and Pallis (2010) traced the influence of local and regional institutional characteristics on port governance models, despite attempts to implement generic governance solutions. In order to explore the process of port development in more detail, Notteboom et al. (2013) applied the concept of institutional plasticity (Strambach, 2010); they argued that, even though port development is path dependent, a port authority can achieve governance reform by a process of adding layers to existing arrangements. The port authority does not, therefore, break from the existing path of development, but instead develops new capabilities and activities through a process described as 'institutional stretching'. This process covers the examples in Chapter 4 of port authorities investing in load centres in the hinterland, and the authors particularly highlighted the importance of informal networking. Jacobs and Notteboom (2011) drew from the economic geography literature to define a movement from 'critical moments' to 'critical junctures' and proposed as a result of their analysis that port authorities have 'windows of opportunity' in which collective action is possible. The authors concluded that 'the question of to what extent critical moments require institutional adaptations in order to materialise into critical junctures needs further thought' (p.1690).

Extending this line of enquiry, the factors that turn critical moments into critical junctures were explored by Wilmsmeier et al. (2013) in an analysis of secondary port development in Latin America and the Caribbean. The theoretical perspective was supplied from the concept of autopoesis (Maturana and Varela, 1980). This

---

1   This section draws on Monios and Wilmsmeier (2013).

concept was originally introduced to port geography by Sanchez and Wilmsmeier (2010), who observed that transport systems exhibit an autopoetic self-organising structure. When a transport system faces pressure from an uncertain environment such as market forces, it seeks to confront challenges to its existing system organisation. If feedback loops are missing then these actions may cause parts of the system to grow in an uncontrollable manner. Finally, through the limitations of its physical characteristics, these actions may lead to an overshooting and collapse of the transport system. This cycle continues so that, with each transformation of the inputs, the system changes its state (Schober, 1991).

The preceding theoretical characterisations of port authorities and terminal operators provide an avenue for future research, exploring such institutional adaptations in more detail. The discussions in this book have reflected how these processes are influenced by globalised norms (e.g. strategy reproductions by global terminal operators) as well as regional specificities. The cyclical process of transport autopoiesis is likely to have an especially high inertia when it comes to changing system variables (Maturana, 1994; Jantsch, 1982), related to the 'lumpiness' of transport infrastructure investment. The factors influencing these system variables, like the factors influencing deconcentration of maritime flows (see Ducruet et al., 2009; Notteboom, 2010; Wilmsmeier and Monios, 2013), will be a mixture of reactive and proactive. More research into these factors and the recursive process of influence is required.

Conclusion and Future Research Agenda

The analysis in the empirical chapters identified several difficulties arising from the nature of intermodal transport and logistics that may challenge successful implementation of port regionalisation strategies. The case studies in chapter 4 showed that ports can actively develop inland terminals, and differences exist between those developed by port authorities and those developed by port terminal operators. Differences can also be observed between those developed by ports and those developed by inland actors. The case of intermodal logistics in Chapter 5 revealed that while rail remains a marginal business, while the industry remains fragmented, while consolidation is not pursued and while fragile government subsidy remains the basis of many flows, intermodal corridors cannot become instruments of hinterland capture and control for ports. The integration processes predicted by the port regionalisation concept cannot happen until the inland logistics system becomes more integrated in comparison to the maritime sector. The institutional analysis in Chapter 6 showed that institutional design often constrains integration between maritime and inland transport systems. The conflict between legitimacy and agency creates barriers and if an infrastructure for collective action is not in place (and it is usually predominately a public infrastructure for collective action), then private firms will not act, thus challenging attempts at port regionalisation and keeping the maritime and inland spaces separate. The multi-scalar formal and informal planning regimes in which each port is situated mean that generic port development strategies based on assumptions of hinterland integration will face several regionally-specific challenges, which concurs with other research on port governance.

While additional case analysis is required, the cases in this book identify potential barriers that may prevent ports controlling or capturing hinterlands through the strategies of integration that the port regionalisation concept suggests. The book also argues for greater disaggregation of the factors that challenge or enable port regionalisation processes, comparing the institutional models of ports and other stakeholders, particularly public sector planners and funders. The institutional analysis in Chapter 7 of governance relationships underpinning operational models goes some way towards achieving this goal. It may be more accurate to state that port regionalisation can only occur as long as a set of favourable commercial and institutional conditions are maintained.

The findings from the cases presented in this book suggest that it is not easy to maintain such conditions, but it has also been shown how they can be transformed. For instance, the commercial conditions can be altered (e.g. port terminals taking a direct role in managing hinterland rail services), as can the institutional conditions (e.g. institutional adaptation to allow port authorities to take direct investments in the hinterland). This best practice can be isolated, analysed and understood through an appreciation of the spatial and institutional characteristics of intermodal transport and logistics as argued in this book.

Recent work on institutional adaptation at ports suggests that port regionalisation, 'imposed on ports', has resulted in a changing institutional design and a transformed relation to their own core competencies, although it is far from clear that this is happening beyond a handful of major ports. Processes of 'institutional plasticity', 'windows of opportunity' and 'autopoeisis' have been identified in the literature, pointing towards a future research agenda examining how the competitive strategies adopted by port authorities and terminal operators fit with the local and regional economic imperatives of landside actors.

References

AAR. (2010). *Class I Railroad Statistics.* Available at: http://www.aar.org/~/media /aar/Industry%20Info/AAR%20Stats%202010%200524.ashx (Accessed 26 October 2010).

AASHTO. (2003). *Transportation: Invest in America; Freight-Rail Bottom Line Report.* Washington, DC: AASHTO.

Abrahamsson, M., Brege, S. (1997). Structural changes in the supply chain. *International Journal of Logistics Management.* 8 (1): 35–44.

Adzigbey, Y., Kunaka, C., Mitiku, T.N. (2007). *Institutional Arrangements for Transport Corridor Management in Sub-Saharan Africa.* SSATP working paper 86. Washington DC: World Bank.

Albrechts, L., Coppens, T. (2003). Megacorridors: striking a balance between the space of flows and the space of places. *Journal of Transport Geography.* 11 (3): 214–24.

Allen, J., Cochrane, A. (2007). Beyond the territorial fix: regional assemblages, politics and power. *Regional Studies.* 41 (9): 1161–75.

Allen, J., Massey, D., Thrift, N. with Charlesworth, J., Court, G., Henry, N., Sarre, P. (1998). *Rethinking the Region.* London: Routledge.

Alphaliner. (2012). *Evolution of carriers fleets.* Available at: http://www.alphaliner. com/liner2/research_files/liner_studies/misc/AlphalinerTopCarriers-2012.pdf (Accessed 25 September 2013).

Amin, A. (1994). The difficult transition from informal economy to Marshallian industrial district. *Area.* 26 (1): 13–24.

Amin, A. (2001). Moving on: institutionalism in economic geography. *Environment & Planning A.* 33 (7): 1237–41.

Amin, A., Thrift, N.J. (1994). Living in the global. In: Amin, A., Thrift, N. (eds). *Globalization, Institutions and Regional Development in Europe.* Oxford: Oxford University Press, pp. 1–22.

Amin, A., Thrift, N.J. (1995). Globalization, institutional "thickness" and the local economy. In: Healey, P., Cameron, S., Davoudi, S., Graham, S., Madinpour, A. (eds). *Managing Cities; The New Urban Context.* Chichester: Wiley, pp. 91–108.

Amin, A., Thrift, N. (2002). *Cities: Reimagining the Urban.* Cambridge: Polity Press.

Aoki, M. (2007). Endogenizing institutions and institutional changes. *Journal of Institutional Economics.* 3 (1): 1–31.

ARC. (2010). *The Heartland Corridor: Opening New Access to Global Opportunity.* Washington, DC: ARC.

Arnold, P., Peeters, D., Thomas, I. (2004). Modelling a rail/road intermodal transportation system. *Transportation Research Part E.* 40 (3): 255–70.

Arthur, W.B. (1994). *Increasing Returns and Path Dependence in the Economy.* Ann Arbor: University of Michigan Press.

Arvis J.F., Carruthers R., Smith G., Willoughby C. (2011). *Connecting Landlocked Developing Countries to Markets: Trade Corridors in the 21st Century.* Washington, DC: World Bank.

Arvis, J.-F., Raballand, G., Marteay, J.-F. (2007). *The Cost of Being Landlocked: Logistics, Costs, and Supply Chain Reliability.* Washington, DC: World Bank.

Baird, A. (2002). Privatization trends at the world's top-100 container ports. *Maritime Policy & Management.* 29 (3): 271–84.

Baird, A.J. (2000). Port privatisation: objectives, extent, process and the UK Experience. *International Journal of Maritime Economics.* 2 (2): 177–94.

Ballis, A., Golias, J. (2002). Comparative evaluation of existing and innovative rail-road freight transport terminals. *Transportation Research Part A.* 36 (7): 593–611.

Baltazar, R., Brooks, M.R. (2001). The governance of port devolution: a tale of two countries. Paper presented at the 9th World Conference on Transport Research, Seoul, 2001.

Barke, M. (1986). *Transport and Trade; Conceptual Frameworks in Geography.* Edinburgh: Oliver & Boyd.

Barney, J. (1991). Firm resources and sustained competitive advantage. *Journal of Management.* 17 (1): 99–120.

Bärthel, F., Woxenius, Y. (2004). Developing intermodal transport for small flows over short distances. *Transportation Planning & Technology.* 27 (5): 403–24.

Beresford, A.K.C., Dubey, R.C. (1991). *Handbook on the Management and Operation of Dry Ports.* RDP/LDC/7. Geneva, Switzerland: UNCTAD.

Beresford, A.K.C., Gardner, B.M., Pettit, S.J., Naniopoulos, A., Wooldridge, C.F. (2004). The UNCTAD and WORKPORT models of port development: evolution or revolution? *Maritime Policy & Management.* 31 (4): 93–107.

Beresford, A., Pettit, S., Xu, Q., Williams, S. (2012). A study of dry port development in China. *Maritime Economics & Logistics.* 14 (1): 73–98.

Beresford, A.K.C. (1999). Modelling freight transport costs: a case study of the UK-Greece corridors. *International Journal of Logistics: Research and Applications.* 2 (3): 229–46.

Bergqvist, R. (2008). Realising logistics opportunities in a public-private collaborative setting: the story of Skaraborg. *Transport Reviews.* 28 (2): 219–37.

Bergqvist, R., Behrends, S. (2011). Assessing the effects of longer vehicles: the case of pre- and post-haulage in intermodal transport chains. *Transport Reviews.* 31 (5): 591–602.

Bergqvist, R., Falkemark, G., Woxenius, J. (2010). Establishing intermodal terminals. *World Review of Intermodal Transportation Research.* 3 (3): 285–302.

Bergqvist, R., Wilmsmeier, G. (2009). Extending the role and concept of dryports: A response to the public consultation of "A sustainable future for transport: Towards an integrated, technology-led and user friendly system" by the Dryport project. Gothenburg: Dryport.

Berkeley, T. (2010). Rail freight in the UK. Presentation given at the International Conference on Intermodal Strategies for Integrating Ports and Hinterlands, Edinburgh, October, 2010.

Bichou, K., Gray, R. (2004). A logistics and supply chain management approach to port performance measurement. *Maritime Policy & Management.* 31 (1): 47–67.

Bichou, K., Gray, R. (2005). A critical review of conventional terminology for classifying seaports. *Transportation Research Part A: Policy and Practice.* 39 (1): 75–92.

Bichou, K. (2009). *Port Operations, Planning and Logistics.* London: Informa Law.

Bird, J. (1963). *The Major Seaports of the United Kingdom.* London: Hutchinson & Co.

Bird, J. (1971). *Seaports and Seaport Terminals.* London: Hutchinson & Co.

Bonacich, E., Wilson, J.B. (2008). *Getting the Goods; Ports, Labour and the Logistics Revolution.* Ithaca, NY: Cornell University Press.

Bontekoning, Y.M., Macharis, C., Trip, J.J. (2004). Is a new applied transportation research field emerging? A review of intermodal rail–truck freight transport literature. *Transportation Research Part A: Policy and Practice.* 38 (1): 1–34.

Bouley, C. (2012). *Manifesto for the 45' palletwide container: a green container for Europe.* Available at: http://issuu.com/cjbouley/docs/manifesto_for_the_45_pallet_wide_container (Accessed 16 March 2012).

Bowen, J. (2008). Moving places: the geography of warehousing in the US. *Journal of Transport Geography.* 16 (6): 379–87.

Bowersox, D.J., Daugherty, P.J., Dröge, C.L., Rogers, D.S., Wardlow, D.L. (1989). *Leading Edge Logistics: Competitive Positioning for the 1990s.* Oak Brook, IL: Council of Logistics Management.

Boyd, J.D. (2010). If you build it … stacking up hopes in the heartland. Journal of Commerce. 6 Sept 2010. Available at: http://www.joc.com/rail-intermodal/if-you-build-it-stacking-hopes-heartland (Accessed 10 April 2011).

Brenner, N. (1999). Beyond state-centrism? Space, territoriality, and geographical scale in globalization studies. *Theory and Society.* 28 (1): 39–78.

Brenner, N. (2004). *New State Spaces; Urban Governance and the Rescaling of Statehood.* Oxford: Oxford UP.

Broeze, F. (2002). *The Globalisation of the Oceans: Containerisation From the 1950s to the Present.* St. Johns, NF, Canada: International Maritime Economic History Association.

Brooks, M.R. (2004). The governance structure of ports. *Review of Network Economics.* 3 (2): 168–83.

Brooks, M.R., Cullinane, K. (eds). (2007). *Devolution, Port Governance and Port Performance.* London: Elsevier.

Brooks, M., Pallis, A.A. (2008). Assessing port governance models: process and performance components. *Maritime Policy & Management*. 35 (4): 411–32.

Bryman, A. (2008). *Social Research Methods*. Oxford: Oxford UP.

Burt, S.L., Sparks, L. (2003). Power and competition in the UK retail grocery market. *British Journal of Management*. 14 (3): 237–54.

Caballini, C., Gattorna, E. (2009). The expansion of the port of Genoa: the Rivalta Scrivia dry port. *UNESCAP Transport and Communications Bulletin for Asia and the Pacific. No. 78: Development of Dry Ports*. New York, UNESCAP.

Callahan, R.F., Pisano, M., Linder, A. (2010). Leadership and strategy: a comparison of the outcomes and institutional designs of the Alameda Corridor and the Alameda Corridor East projects. *Public Works Management & Policy*. 14 (3): 263–87.

Castells, M. (1996). *The Rise of the Network Society*. The Information Age: Economy, Society and Culture, vol. 1. Oxford: Blackwell.

Chapman, D., Pratt, D., Larkham, P., Dickins, I. (2003). Concepts and definitions of corridors: evidence from England's Midlands. *Journal of Transport Geography*. 11 (3): 179–91.

Charlier, J.J., Ridolfi, G. (1994). Intermodal transportation in Europe: of modes, corridors and nodes. *Maritime Policy & Management*. 21 (3): 237–50.

Chatterjee, L., Lakshmanan, T.R. (2008). Intermodal freight transport in the United States. In: Konings, R., Priemus, H., Nijkamp, P. (eds). *The Future of Intermodal Freight Transport*. Cheltenham: Edward Elgar, pp. 34–57.

Chen, H., Daugherty, P.J., Roath, A.S. (2009). Defining and operationalizing supply chain process integration. *Journal of Business Logistics*. 30 (1): 63–84.

Choong, S.T., Cole, M.H., Kutanoglu, E. (2002). Empty container management for intermodal transportation networks. *Transportation Research Part E: Logistics and Transportation Review*. 38 (6): 423–38.

Christaller, W. (1933). *Die Zentralen Orte in Süddeutschland (Central Places in Southern Germany)*. Trans. C.W. Baskin (1966). Englewood Cliffs, NJ: Prentice Hall.

Christopher M. (1992). *Logistics and Supply Chain Management: Strategies for Reducing Costs and Improving Services*. London: Pitman Publishing.

Cidell, J. (2010). Concentration and decentralization: the new geography of freight distribution in US metropolitan areas. *Journal of Transport Geography*. 18 (3): 363–71.

Cidell, J. (2013). From hinterland to distribution centre: the Chicago region's shifting gateway function. In: Hall, P.V., Hesse, M. (eds). *Cities, Regions and Flows*. Abingdon: Routledge, pp. 114–28.

Coase, R.H. (1937). The nature of the firm. *Economica*. 4 (16): 386–405.

Coase, R.H. (1983). The new institutional economics. *Journal of Institutional and Theoretical Economics*. 140 (1): 229–31.

Containerisation International. (2012). Available at: http://www.ci-online.co.uk/default.asp (Accessed 2 September 2013).

Cooke, P., Morgan, K. (1998). *The Associational Economy: Firms, Regions and Innovation*. Oxford: Oxford UP.

Coulson, A., Ferrario, C. (2007). 'Institutional thickness': local governance and economic development in Birmingham, England. *International Journal of Urban and Regional Research*. 31 (3): 591–615.

CREATE. (2005). *CREATE Final Feasibility Plan*. Chicago: CREATE.

Cronbach, L. (1975). Beyond the two disciplines of scientific psychology. *American Psychology.* 30 (2): 116–27.

Cruijssen, F., Dullaert, W., Fleuren, H. (2007). Horizontal cooperation in transport and logistics: a literature review. *Transportation Journal.* 46: 22–39.

Cullinane, K.P.B., Wilmsmeier, G. (2011). The Contribution of the Dry Port Concept to the Extension of Port Life Cycles. In: Böse, J.W. (ed.). *Handbook of Terminal Planning*. New York, Springer, pp. 359–80.

Cullinane, K.P.B., Wang, Y. (2012). The hierarchical configuration of the container port industry: an application of multiple linkage analysis. *Maritime Policy & Management.* 39 (2): 169–87.

Cullinane, K., Song, D.W. (2002). Port privatisation policy and practice. *Maritime Policy and Management.* 22 (1): 55–75.

Cullinane, K.P.B., Khanna, M. (1999). Economies of scale in large container ships. *Journal of Transport Economics and Policy.* 33 (2): 185–208.

Curtis, C., Lowe, N. (2012). *Institutional Barriers to Sustainable Transport*. Farnham, Surrey: Ashgate.

Dablanc, L., Ross, C. (2012). Atlanta: a mega logistics center in the Piedmont Atlantic Megaregion (PAM). *Journal of Transport Geography.* 24: 432–42.

David, P.A. (1985). Clio and the Economics of QWERTY. *American Economic Review.* 75: 332–7.

De Langen, P.W. (2008). *Ensuring Hinterland Access: the Role of Port Authorities*. JTRC OECD/ITF Discussion Paper 2008–11. Paris: ITF.

De Langen, P.W. (2004). *The performance of seaport clusters, a framework to analyze cluster performance and an application to the seaport clusters of Durban, Rotterdam and the Lower Mississippi*. Rotterdam: ERIM PhD series.

De Langen, P.W., Chouly, A. (2004). Hinterland access regimes in seaports. *European Journal of Transport and Infrastructure Research.* 4 (4): 361–80.

De Langen, P., Visser, E.-J. (2005). Collective action regimes in seaport clusters: the case of the Lower Mississippi port cluster. *Journal of Transport Geography.* 13 (2): 173–86.

De Vries, J., Priemus, H. (2003). Megacorridors in north-west Europe: issues for transnational spatial governance. *Journal of Transport Geography.* 11 (3): 225–33.

de Wulf, L., Sokol, J. (eds). (2005). *Customs Modernization Handbook*. Washington, DC: The World Bank.

Debrie, J., Gouvernal, E., Slack, B. (2007). Port devolution revisited: the case of regional ports and the role of lower tier governments. *Journal of Transport Geography.* 15 (6): 455–64.

Debrie, J., Lavaud-Letilleul, V., Parola, F. (2013). Shaping port governance: the territorial trajectories of reform. *Journal of Transport Geography*. 27: 56–65.

Dekker, R., van Asperen, E., Ochtman, G., Kusters, W. (2009). Floating stocks in FMCG supply chains: using intermodal transport to facilitate advance deployment. *International Journal of Physical Distribution & Logistics Management*. 39 (8): 632–48.

Department for Transport. (2011). *DfT Port Statistics*. London: DfT.

Department for Transport. (2011). *Britain's Transport Infrastructure. Strategic Rail Freight Network: The Longer Term Vision*. London: DfT.

Dicken, P. (2011). *Global Shift*. 6th edn. New York: Guildford.

Djankov, S., Freud, C., Pham, C.C. (2005). *Trading on Time*. Research paper 3909. Washington DC: World Bank.

Drewry Shipping Consultants. (2012). *Global Container Terminal Operators Annual Review and Forecast 2012*. London: Drewry Publishing.

Drewry Shipping Consultants. (2013). *Container Market – 2012/13. Annual Review and Forecast*. London: Drewry publishing.

Ducruet, C. (2009). Port regions and globalization. In: Notteboom, T., Ducruet, D., deLangen, P. (eds). *Ports in Proximity*. Farnham: Ashgate, pp. 41–54.

Ducruet, C., Lee, S.W. (2006). Frontline soldiers of globalisation: Port-city evolution and regional competition. *GeoJournal*. 67 (2): 107–22.

Ducruet, C., Roussin, S., Jo, J.-C. (2009). Going west? Spatial polarization of the North Korean port system. *Journal of Transport Geography*. 17 (5): 357–68.

Ducruet, C., Van der Horst, M. (2009). Transport integration at European ports: measuring the role and position of intermediaries. *EJTIR*. 9 (2): 121–42.

Dussauge, P., Garrette, B. (1997). Anticipating the evolutions and outcomes of strategic alliances between rival firms. *International Studies of Management & Organization*. 27: 104–26.

Dyer, J.H., Singh, H. (1998). The relational view: cooperative strategy and sources of interorganizational competitive advantage. *Academy of Management Review*. 23 (4): 660–79.

Eng-Larsson, F., Kohn, C. (2012). Modal shift for greener logistics – the shipper's perspective. *International Journal of Physical Distribution and Logistics Management*. 42 (1): 36–59.

European Commission. (2001). *European Transport Policy for 2010: Time to Decide*. Luxembourg: European Commission.

Evangelista, P., Morvillo, A. (2000). Maritime transport in the Italian logistic system. *Maritime Policy & Management*. 27 (4): 335–52.

Everett, S., Robinson, R. (1998). Port reform in Australia: issues in the ownership debate. *Maritime Policy & Management*. 25 (4): 41–62.

Fan, L., Wilson, W.W., Tolliver, D. (2009). Logistical rivalries and port competition for container flows to US markets: impacts of changes in Canada's logistics system and expansion of the Panama Canal. *Maritime Economics & Logistics*. 11 (4): 327–57.

Fawcett, S.E., Magnan, G.M., McCarter, M.W. (2008a). A three-stage implementation model for supply chain collaboration. *Journal of Business Logistics*. 29 (1): 93–112.

Fawcett, S.E., Magnan, G.M., McCarter, M.W. (2008b). Benefits, barriers and bridges to effective supply chain management. *Supply Chain Management: An International Journal*. 13 (1): 35–48.

FDT. (2007). *Feasibility Study on the Network Operation of Hinterland Hubs (Dry Port Concept) to Improve and Modernise Ports' Connections to the Hinterland and to Improve Networking*. Aalborg, Denmark: FDT.

FDT. (2009). *The Dry Port: Concepts and Perspectives*. Aalborg, Denmark: FDT.

Fernie, J., McKinnon, A.C. (2003). The Grocery Supply Chain in the UK: Improving Efficiency in the Logistics Network. *International Review of Retail, Distribution and Consumer Research*. 13 (2): 161–74.

Fernie, J., Pfab, F., Merchant, C. (2000). Retail grocery logistics in the UK. *International Journal of Logistics Management*. 11 (2): 83–90.

Fernie, J., Sparks, L., McKinnon, A.C. (2010). Retail logistics in the UK: past, present and future. *International Journal of Retail and Distribution Management*. 38 (11/12): 894–914.

Ferrari, C., Musso, E. (2011). Italian ports: towards a new governance? *Maritime Policy & Management*. 38 (3): 335–46.

Ferreira, L., Sigut, J. (1993). Measuring the performance of intermodal freight terminals. *Transportation Planning & Technology*. 17 (3): 269–80.

FHA. (2010). *Freight Analysis Framework, version 3.1, 2010*. Washington DC: U.S. Department of Transportation, Federal Highway Administration, Office of Freight Management and Operations.

Flämig, H., Hesse, M. (2011). Placing dryports. Port regionalization as a planning challenge – the case of Hamburg, Germany, and the Süderelbe. *Research in Transportation Economics*. 33 (1): 42–50.

Fleming, D.K., Hayuth, Y. (1994). Spatial characteristics of transportation hubs: centrality and intermediacy. *Journal of Transport Geography*. 2 (1): 3–18.

Flyvbjerg, B. (2006). Five misunderstandings about case-study research. *Qualitative Inquiry*. 12 (2): 219–45.

Forum for the Future. (2007). *Retail Futures; Scenarios for the Future of UK Retail and Sustainable Development*. London: Forum for the Future.

FRA. (2012). FRA website. Available at: http://www.fra.dot.gov/rpd/freight/1486.shtml (Accessed 4 November 2010).

Frémont, A., Franc, P. (2010). Hinterland transportation in Europe: combined transport versus road transport. *Journal of Transport Geography*. 18 (4): 548–56.

Frémont, A., Soppé, M. (2007). Northern European range: shipping line concentration and port hierarchy. In: Wang, J., Olivier, D., Notteboom, T., Slack, B. (eds). *Ports, Cities and Global Supply Chains*. Aldershot: Ashgate, pp. 105–20.

Friedman, T.L. (2005). *The World is Flat: A Brief History of the Twenty-First Century*. New York: Farrar, Straus and Giroux.

FTA. (2012). *On Track: Retailers Using Rail Freight to Make Cost and Carbon Savings*. London: FTA.

Fundación Valenciaport. (2010). Personal communication, 13 October, 2010.

Gangwar, R., Morris, S., Pandey, A., Raghuram, G. (2012). Container movement by rail in India: a review of policy evolution. *Transport Policy*. 22 (1): 20–28.

Garnwa, P., Beresford, A., Pettit, S. (2009). Dry ports: a comparative study of the United Kingdom and Nigeria. In: *Transport and Communications Bulletin for Asia and the Pacific No. 78: Development of Dry Ports*. New York: UNESCAP.

Geerlings, H., Stead, D. (2003). The integration of land use planning, transport and environment in European policy and research. *Transport Policy*. 10 (3): 187–96.

Gifford, J.L., Stalebrink, O.J. (2002). Remaking transportation organizations for the 21st century: consortia and the value of organizational learning. *Transportation Research Part A*. 36 (7): 645–57.

Gimenez, C., Ventura, E. (2005). Logistics-production, logistics-marketing and external integration. *International Journal of Operations & Production Management*. 25 (1): 20–38.

Golicic, S.L., Mentzer, J.T. (2006). An empirical examination of relationship magnitude. *Journal of Business Logistics*. 27 (1): 81–108.

González, S., Healey, P. (2005). A sociological institutionalist approach to the study of innovation in governance capacity. *Urban Studies*. 42 (11): 2055–69.

Goodwin, A. (2010). *The Alameda Corridor: A Project of National Significance*. Carson, CA: ACTA.

Goodwin, M., Jones, M., Jones, R. (2005). Devolution, constitutional change and economic development: explaining and understanding the new institutional geographies of the British state. *Regional Studies*. 39 (4): 421–36.

Gouvernal, E., Debrie, J., Slack, B. (2005). Dynamics of change in the port system of the western Mediterranean. *Maritime Policy & Management*. 32 (2): 107–21.

Graham, M.G. (1998). Stability and competition in intermodal container shipping: finding a balance. *Maritime Policy & Management*. 25 (2): 129–47.

Grawe, S.J., Daugherty, P.J., Dant, R.P. (2012). Logistics service providers and their customers: gaining commitment through organizational implants. *Journal of Business Logistics*. 33 (1): 50–63.

Groenewegen, J., De Long, M. (2008). Assessing the potential of new institutional economics to explain institutional change: the case of road management liberalization in the Nordic countries. *Journal of Institutional Economics*. 4 (1): 51–71.

Groothedde, B., Ruijgrok, C., Tavasszy, L. (2005). Towards collaborative, intermodal hub networks: A case study in the fast moving consumer goods market. *Transportation Research Part E: Logistics and Transportation Review*. 41 (6): 567–83.

Guan, W., Rehme, J. (2012). Vertical integration in supply chains: driving forces and consequences for a manufacturer's downstream integration. *Supply Chain Management: An International Journal.* 17 (2): 187–201.

Hailey, R. (2011). CMA CGM signs two-year rail deal with DB Schenker at Southampton. Lloyd's List. 18 April 2011. Available at: http://www.lloyds list.com/ll/sector/ports-and-logistics/article368643.ece (Accessed 25 August 2013).

Hall, D. (2010). Transport geography and new European realities: a critique. *Journal of Transport Geography.* 18 (1): 1–13.

Hall, P.V. (2003). Regional institutional convergence? Reflections from the Baltimore waterfront. *Economic Geography.* 79 (4): 347–63.

Hall, P.V., Hesse, M. (2013). Reconciling cities and flows in geography and regional studies. In: Hall, P.V., Hesse, M. (eds). *Cities, Regions and Flows.* Abingdon: Routledge, pp. 3–20.

Hall, P.V., Jacobs, W. (2012). Why are maritime ports (still) urban, and why should policy-makers care? *Maritime Policy & Management.* 39 (2): 189–206.

Hall, P.V., Jacobs, W. (2010). Shifting proximities: the maritime ports sector in an era of global supply chains. *Regional Studies.* 44 (9): 1103–15.

Hall, P., Hesse, M., Rodrigue, J.-P. (2006). Reexploring the interface between economic and transport geography. *Environment & Planning A.* 38 (7): 1401–8.

Hall, P., McCalla, R.J., Comtois, C., Slack, B. (eds) (2011). *Integrating Seaports and Trade Corridors.* Aldershot: Ashgate.

Halldórsson, A., Skjøtt-Larsen, T. (2006). Dynamics of relationship governance in TPL arrangements – a dyadic perspective. *International Journal of Physical Distribution & Logistics Management.* 36 (7): 490–506.

Hammersley, M. (1992). *What's Wrong with Ethnography?* London: Routledge.

Hanaoka, S., Regmi, M.B. (2011). Promoting intermodal freight transport through the development of dry ports in Asia: an environmental perspective. *IATSS Research.* 35 (1): 16–23.

Haynes, K.E., Gifford, J.L., Pelletiere, D. (2005). Sustainable transportation institutions and regional evolution: global and local perspectives. *Journal of Transport Geography.* 13 (3): 207–21.

Hayuth, Y. (1980). Inland container terminal – function and rationale. *Maritime Policy and Management.* 7 (4): 283–9.

Hayuth, Y. (1981). Containerization and the load center concept. *Economic Geography.* 57 (2): 160–76.

Hayuth, Y. (2007). Globalisation and the port-urban interface: conflicts and opportunities. In: Wang, J., Olivier, D., Notteboom, T., Slack, B. (eds). *Ports, Cities and Global Supply Chains.* Aldershot: Ashgate, pp. 141–56.

Haywood, R. (1999). Land development implications of the British rail freight renaissance. *Journal of Transport Geography.* 7 (4): 263–75.

Heaver, T., Meersman, H., Moglia, F., Van de Voorde, E. (2000). Do mergers and alliances influence European shipping and port competition? *Maritime Policy & Management.* 27 (4): 363–73.

Heaver, T., Meersman, H., Van de Voorde, E. (2001). Co-operation and competition in international container transport: strategies for ports. *Maritime Policy & Management*. 28 (3): 293–305.

Henry, N., Pinch, S. (2001). Neo-Marshallian nodes, institutional thickness, and Britain's 'Motor Sport Valley': thick or thin? *Environment & Planning A.* 33 (7): 1169–83.

Hernández-Espallardo, M., Arcas-Lario, N. (2003). The effects of authoritative mechanisms of coordination on market orientation in asymmetrical channel partnerships. *International Journal of Research in Marketing.* 20 (2): 133–52.

Hernández-Espallardo, M., Rodríguez-Orejuela, A., Sánchez-Pérez, M. (2010). Inter-organizational governance, learning and performance in supply chains. *Supply Chain Management: An International Journal.* 15 (2): 101–14.

Hesse, M. (2004). Land for Logistics. locational dynamics, real estate markets and political regulation of regional distribution complexes. *Tijdschrift voor Sociale en Economische Geografie.* 95 (2): 162–73.

Hesse, M. (2008). *The City as a Terminal. Logistics and Freight Distribution in an Urban Context.* Aldershot: Ashgate.

Hesse, M. (2013). Cities and flows: re-asserting a relationship as fundamental as it is delicate. *Journal of Transport Geography.* 29: 33–42.

Hesse, M., Rodrigue, J.-P. (2004). The transport geography of logistics and freight distribution. *Journal of Transport Geography.* 12 (3): 171–84.

Hingley, M, Lindgreen, A., Grant, D.B., Kane, C. (2011). Using fourth-party logistics management to improve horizontal collaboration among grocery retailers. *Supply Chain Management: An International Journal.* 16 (5): 316–27.

Hoare, A.G. (1986). British ports and their export hinterlands: a rapidly changing geography. *Geografiska Annaler.* 68B (1): 29–40.

Hoffmann, J. (2001). Latin American ports: results and determinants of private sector participation. *International Journal of Maritime Economics.* 3 (2): 221–41.

Holguin-Veras, J., Paaswell, R., Perl, A. (2008). The role of government in fostering intermodal transport innovations: perceived lessons and obstacles in the United States. In: Konings, R., Priemus, H., Nijkamp, P. (eds). *The Future of Intermodal Freight Transport.* Cheltenham: Edward Elgar, pp. 302–24.

Höltgen, D. (1996). *Intermodal Terminals in the Trans-European Network.* Discussion Paper. Rotterdam: European Centre for Infrastructure Studies.

Hooghe, L., Marks, G. (2003). Unraveling the central state, but how? Types of multi-level governance. *American Political Science Review.* 97 (2): 233–43.

Hooghe, L., Marks, G. (2001). *Multi-Level Governance and European Integration.* Boulder, Col.: Rowman & Littlefield.

Hotelling, H. (1929). Stability in competition. *Economic Journal.* 39 (153): 41–57.

Hoyle, B.S. (1968). East African seaports: an application of the concept of 'anyport'. *Transactions & Papers of the Institute of British Geographers.* 44: 163–83.

Hoyle, B.S. (2000). Global and local change on the port-city waterfront. *Geographical Review.* 90 (3): 395–417.

Humphries, A.S., Towriss, J., Wilding, R. (2007). A taxonomy of highly interdependent supply chain relationships. *The International Journal of Logistics Management*. 18 (3): 385–401.

Iannone, F. (2012). Innovation in port-hinterland connections. The case of the Campanian logistic system in Southern Italy. *Maritime Economics & Logistics*. 14 (1): 33–72.

IGD. (2012). Over 200 million food miles removed from UK roads. Available at http://www.igd.com/print.asp?pid=1&pflid=6&plid=5&pcid=2303 (Accessed 11 April 2012).

Jaccoby, S.M. (1990). The new institutionalism: what can it learn from the old? *Industrial Relations*. 29 (2): 316–59.

Jacobs, W. (2007). Port competition between Los Angeles and Long Beach: an institutional analysis. *Tijdschrift voor Economische en Sociale Geografie*. 98 (3): 360–72.

Jacobs, W., Notteboom, T. (2011). An evolutionary perspective on regional port systems: the role of windows of opportunity in shaping seaport competition. *Environment & Planning A*. 43 (7): 1674–92.

Janic, M. (2007). Modelling the full costs of an intermodal and road freight transport network. *Transportation Research Part D: Transport and Environment*. 12 (1): 33–44.

Jantsch, E. (1982). *Die Selbstorganisation des Universums*. Munich: Hanser.

Jarzemskis, A., Vasiliauskas, A.V. (2007). Research on dry port concept as intermodal node. *Transport*. 22 (3): 207–13.

Jessop, B. (2001). Institutional (re)turns and the strategic-relational approach. *Environment & Planning A*. 33 (7): 1213–35.

Jessop, B. (1990). *State Theory: Putting Capitalist States in their Place*. Cambridge: Polity.

Jones, M. (1997). Spatial selectivity of the state? The regulationist enigma and local struggles over economic governance. *Environment and Planning A*. 29 (5): 831–64.

Jones, P., Comfort, D., Hillier, D. (2005). Corporate social responsibility and the UK's top ten retailers. *International Journal of Retail and Distribution Management*. 33 (12): 882–92.

Jones, P., Comfort, D., Hillier, D. (2008). UK retailing through the looking glass. *International Journal of Retail and Distribution Management*. 36 (7): 564–70.

Jordan, A., Wurzel, R.K., Zito, A. (2005). The rise of 'new' policy instruments in comparative perspective: has governance eclipsed government? *Political Studies*. 53 (3): 477–96.

Kelle, U. (1997). Theory Building in Qualitative Research and Computer Programs for the Management of Textual Data. *Sociological Research Online*. 2 (2). Available at: www.socresonline.org.uk/2/2/1.html. (Accessed 19 September 2013).

Kim, N.S., Wee, B.V. (2011). The relative importance of factors that influence the break-even distance of intermodal freight transport systems. *Journal of Transport Geography.* 19 (4): 859–75.

Klint, M.B., Sjöberg, U. (2003). Towards a comprehensive SCP-model for analysing strategic networks/alliances. *International Journal of Physical Distribution & Logistics Management.* 33 (5): 408–26.

Knowles, R., Shaw, J., Docherty, I. (2008). *Transport Geographies: Mobilities, Flows and Apaces.* Oxford: Blackwell.

Konings, J.W. (1996). Integrated centres for the transhipment, storage, collection and distribution of goods. *Transport Policy.* 3 (1): 3–11.

Konings, R. (2007). Opportunities to improve container barge handling in the port of Rotterdam from a transport network perspective. *Journal of Transport Geography.* 15 (6): 443–54.

Konings, R., Kreutzberger, E., Maraš, V. (2013). Major considerations in developing a hub-and-spoke network to improve the cost performance of container barge transport in the hinterland: the case of the port of Rotterdam. *Journal of Transport Geography.* 29: 63–73.

Konings, R., Priemus, H., Nijkamp, P. (eds). (2008). *The Future of Intermodal Freight Transport.* Cheltenham: Edward Elgar.

Kreutzberger, E.D. (2008). Distance and time in intermodal goods transport networks in Europe: a generic approach. *Transportation Research Part A: Policy & Practice.* 42 (7): 973–93.

Kuipers, B. (2002). The rise and fall of the maritime mainport. Paper presented at the European Transport Conference, Cambridge, September 2002.

Kunaka, C. (2013). Dry ports and trade logistics in Africa. In: Bergqvist, R., Cullinane, K.P.B., Wilmsmeier, G. (eds). *Dry Ports: A Global Perspective.* London: Ashgate, pp. 83–105.

Lambert, D.M., Emmelhainz, M.A., Gardner, J.T. (1999). Building successful logistics partnerships. *Journal of Business Logistics.* 20 (1): 165–81.

Lambert, D.M., García-Dastugue, S.J., Croxton, K.L. (2008). The role of logistics managers in the cross-functional implementation of supply chain management. *Journal of Business Logistics.* 29 (1): 113–32.

Lammgård, C. (2012). Intermodal train services: a business challenge and a measure for decarbonisation for logistics service providers. *Research in Transportation Business & Management.* 5: 48–56.

Lavie, D. (2006). The competitive advantage of interconnected firms: an extension of the resource-based view. *Academy of Management Review.* 31 (3): 638–58.

le Blanc, H.M., Cruijssen, F., Fleuren, H.A., de Koster, M.B.M. (2006). Factory gate pricing: An analysis of the Dutch retail distribution. *European Journal of Operational Research.* 174 (3): 1950–67.

LeCompte, M.D., Goetz, J.P. (1982). Problems of reliability and validity in ethnographic research. *Review of Educational Research.* 52 (2): 31–60.

Lee, S.-W., Song, D.-W., Ducruet, C. (2008). A tale of Asia's world ports: the spatial evolution in global hub port cities. *Geoforum.* 39 (1): 372–85.

Legacy, C., Curtis, A., Sturup, S. (2012). Is there a good governance model for the delivery of contemporary transport policy and practice? An examination of Melbourne and Perth. *Transport Policy.* 19 (1): 8–16.

Lehtinan, J., Bask, A.H. (2012). Analysis of business models for potential 3Mode transport corridor. *Journal of Transport Geography.* 22 (1): 96–108.

Lemoine, O.W., Skjoett-Larsen, T. (2004). Reconfigurations of supply chains and implications for transport; a Danish study. *International Journal of Physical Distribution and Logistics Management.* 34 (10): 793–810.

Levinson, M. (2006). *The Box: How the Shipping Container Made the World Smaller and the World Economy Bigger.* Princeton: Princeton UP.

Lewis, J., Ritchie, J. (2003). Generalising from qualitative research. In: Ritchie, J., Lewis, J. (eds). *Qualitative Research Practice.* London: Sage, pp. 263–86.

Liedtke, G., Carrillo Murillo, D.G. (2012). Assessment of policy strategies to develop intermodal services: the case of inland terminals in Germany. *Transport Policy.* 24 (C): 168–78.

Limbourg, S., Jourquin, B. (2009). Optimal rail-road container terminal locations on the European network. *Transportation Research Part E.* 45 (4): 551–63.

Lipietz, A. (1994). The national and the regional: their autonomy vis-à-vis the capitalist world crisis. In: Palan, R., Gills, B. (eds). *Transcending the State-Global Divide: A Neo-Structuralist Agenda in International Relations.* London: Lynne Reimer, pp. 23–43.

Lösch, A. (1940). *Die Räumliche Ordnung der Wirtschaft (The Economics of Location).* Trans. W.W. Woglom, W.F. Stolper (1954). New Haven, CT: Yale UP.

Lovering, J. (1999). Theory led by policy: the inadequacies of the 'new regionalism' (illustrated from the case of Wales). *International Journal of Urban and Regional Research.* 23 (2): 379–95.

Lowe, D. (2005). *Intermodal Freight Transport.* Oxford: Elsevier Butterworth-Heinemann.

Lu, J., Chang, Z. (2013). The construction of seamless supply chain network – development of "dry ports" in China. In: Bergqvist, R., Cullinane, K.P.B., Wilmsmeier, G. (eds). *Dry Ports: A Global Perspective.* London: Ashgate, pp. 155–69.

Macharis, C., Pekin, E. (2009). Assessing policy measures for the stimulation of intermodal transport: a GIS-based policy analysis. *Journal of Transport Geography.* 17 (6): 500–508.

MacLeod, G. (1997). 'Institutional thickness' and industrial governance in Lowland Scotland. *Area.* 29 (4): 299–311.

MacLeod, G. (2001). Beyond soft institutionalism: accumulation, regulation and their geographical fixes. *Environment & Planning A.* 33 (7): 1145–67.

Mangan, J., Lalwani, C., Fynes, B. (2008). Port-centric logistics. *The International Journal of Logistics Management.* 19 (1): 29–41.

Mangan, J., Lalwani, C., Gardner, B. (2004). Combining quantitative and qualitative methodologies in logistics research. *International Journal of Physical Distribution and Logistics Management.* 34 (7): 565–78.

Marks, G. (1993). Structural Policy and Multilevel Governance in the EC. In: Cafruny, A., Rosenthal, G. (eds). *The State of the European Community*. Boulder: Lynne Rienner, pp. 391–411.

Markusen, A. (2003). Fuzzy concepts, scanty evidence, policy distance: the case for rigour and policy relevance in critical regional studies. *Regional Studies*. 37 (6–7): 701–17.

Marsden, G., Rye, T. (2010). The governance of transport and climate change. *Journal of Transport Geography*. 18 (6): 669–78.

Martí-Henneberg, J. (2013). European integration and national models for railway networks (1840–2010). *Journal of Transport Geography*. 26: 126–38.

Martin, C. (2013). Shipping container mobilities, seamless compatibility and the global surface of logistical integration. *Environment & Planning A*. 45 (5): 1021–36.

Martin, R. (2000). Institutional approaches in economic geography. In: Sheppard, E., Barnes, T.J. (eds). *A Companion to Economic Geography*. Malden: Blackwell, pp. 77–94.

Mason, R., Lalwani, C., Boughton, R. (2007). Combining vertical and horizontal collaboration for transport optimisation. *Supply Chain Management: An International Journal*. 12 (3): 187–99.

Maturana, H.R. (1994). *Was ist Erkennen?* Munich: Piper.

Maturana, H.R., Varela, F.J. (1980). *Autopoiesis and Cognition*. Dordrecht, Holland: D. Reidel.

McCalla, R.J. (1999). Global change, local pain: intermodal seaport terminals and their service areas. *Journal of Transport Geography*. 7 (4): 247–54.

McCalla, R.J. (2009). Gateways are more than ports: the Canadian example of cooperation among stakeholders. In: Notteboom, T., Ducruet, D., deLangen, P. (eds). *Ports in Proximity*. Farnham: Ashgate, pp. 115–31.

McCalla, R.J., Slack, B., Comtois, C. (2001). Intermodal freight terminals: locality and industrial linkages. *The Canadian Geographer*. 45 (3): 404–13.

McCalla, R.J., Slack, B., Comtois, C. (2004). Dealing with globalisation at the regional and local level: the case of contemporary containerization. *The Canadian Geographer*. 48 (4): 473–87.

McCrary, T.P. (2010). *National Gateway Update*. Jacksonville, FL: CSX.

McKinnon, A. (2009). The present and future land requirements of logistical activities. *Land Use Policy*. 26 (S): S293-S301.

McKinnon, A. (2010). *Britain Without Double-deck Lorries*. Edinburgh: Heriot-Watt University.

McKinnon, A., Edwards, J. (2012). Opportunities for improving vehicle utilisation. In: McKinnon, A., Browne, M., Whiteing, A. (eds). *Green Logistics; Improving the Environmental Sustainability of Logistics*. 2nd edn. London: KoganPage, pp. 205–22.

MDS Transmodal Ltd. (2002). *Opportunities for developing sustainable freight facilities in Scotland*. Report prepared for the Scottish Executive. Edinburgh: Scottish Executive.

Mentzner, J.T., Min, S., Bobbitt, L.M. (2004). Toward a unified theory of logistics. *International Journal of Physical Distribution and Logistics Management.* 34 (8): 606–27.

Merriam, S.B. (1988). *Case Study Research in Education: A Qualitative Approach.* San Francisco: Jossey-Bass.

Meyer, J.W., Rowan, B. (1977). Institutionalized organizations: formal structure as myth and ceremony. *American Journal of Sociology.* 83 (2): 340–63.

Meyer, J.W., Scott, R.S. (1983). Centralization and the legitimacy problems of local government. In: Meyer, J.W., Scott, R.S. (eds). *Organizational Environments: Ritual and Rationality.* Beverly Hills, CA: Sage, pp. 199–215.

Miles, M.B., Huberman, A.M. (1994). *Qualitative Data Analysis.* Thousand Oaks, CA: Sage.

Min, S., Roath, A.S., Daugherty, P.J., Genchev, S.E., Chen, H., Arndt, A.D., Richey, R.G. (2005). Supply chain collaboration: what's happening? *The International Journal of Logistics Management.* 16 (2): 237–56.

Moe, T.M. (1990). Political institutions: the neglected side of the story. *Journal of Law, Economics and Organization.* 6 (special issue): 213–53.

Moglia, F., Sanguineri, M. (2003). Port planning: the need for a new approach? *Maritime Economics & Logistics.* 5 (4): 413–25.

Monios, J. (2011). The role of inland terminal development in the hinterland access strategies of Spanish ports. *Research in Transportation Economics.* 33 (1): 59–66.

Monios, J. (2014). Intermodal transport as a regional development strategy: the case of Italian freight villages. *Growth and Change.* In press.

Monios, J., Lambert, B. (2013a). The heartland intermodal corridor: public-private partnerships and the transformation of institutional settings. *Journal of Transport Geography.* 27: 36–45.

Monios, J., Lambert, B. (2013b). Intermodal freight corridor development in the United States. In: Bergqvist, R., Cullinane, K.P.B., Wilmsmeier, G. (eds). *Dry Ports: A Global Perspective.* London: Ashgate, pp. 197–218.

Monios, J., Wang, Y. (2013). Spatial and institutional characteristics of inland port development in China. *GeoJournal.* 78 (5): 897–913.

Monios, J., Wilmsmeier, G. (2012a). Giving a direction to port regionalisation. *Transportation Research Part A: Policy & Practice.* 46 (10): 1551–61.

Monios, J., Wilmsmeier, G. (2012b). Port-centric logistics, dry ports and offshore logistics hubs: strategies to overcome double peripherality? *Maritime Policy & Management.* 39 (2): 207–26.

Monios, J., Wilmsmeier, G. (2013). The role of intermodal transport in port regionalisation. *Transport Policy.* 30: 161–72.

Näslund, D. (2002). Logistics needs qualitative research – especially action research. *International Journal of Physical Distribution and Logistics Management.* 32 (5): 321–38.

Nathan Associates. (2011). *Corridor Diagnostic Study of the Northern and Central Corridors of East Africa. Action Plan.* Volume 1: main report. Arusha, Tanzania: East African Community.

National Audit Office. (1996). *Department of Transport: Freight Facilities Grants in England*. London: NAO.

Network Rail. (2007). *Freight Route Utilisation Strategy*. London: Network Rail.

Network Rail. (2011). *Initial Industry Plan (Scotland)*. *Proposals for Control Period 5*. London: Network Rail.

Ng, A.K.Y. (2013). The evolution and research trends of port geography. *The Professional Geographer.* 65 (1): 65–86.

Ng, K.Y.A., Cetin, I.B. (2012). Locational characteristics of dry ports in developing economies: some lessons from Northern India. *Regional Studies.* 46 (6): 757–73.

Ng, K.Y.A., Gujar, G.C. (2009a). The spatial characteristics of inland transport hubs: evidences from Southern India. *Journal of Transport Geography.* 17 (5): 346–56.

Ng, K.Y.A., Gujar, G.C. (2009b). Government policies, efficiency and competitiveness: the case of dry ports in India. *Transport Policy.* 16 (5): 232–9.

Ng, A.K.Y., Padilha, F., Pallis, A.A. (2013). Institutions, bureaucratic and logistical roles of dry ports: the Brazilian experience. *Journal of Transport Geography.* 27 (1): 46–55.

Ng, A.K.Y., Pallis, A.A. (2010). Port governance reforms in diversified institutional frameworks: generic solutions, implementation asymmetries. *Environment & Planning A.* 42 (9): 2147–67.

Ng, A.K.Y, Tongzon, J.L. (2010). The transportation sector of India's economy: dry ports as catalysts for regional development. *Eurasian Geography & Economics.* 51 (5): 1–14.

North, D.C. (1990). *Institutions, Institutional Change and Economic Performance*. Cambridge: Cambridge University Press.

Notteboom, T.E. (1997). Concentration and load centre development in the European container port system. *Journal of Transport Geography.* 5 (2): 99–115.

Notteboom, T.E. (2005). The peripheral port challenge in container port systems. In: Leggate, H., McConville, J., Morvillo, A. (eds). *International Maritime Transport: Perspectives*. London: Routledge, pp. 173–88.

Notteboom, T. (2007). The changing face of the terminal operator business: lessons for the regulator. Paper presented at the ACCC Regulatory Conference. Gold Coast, Australia, July 2007.

Notteboom, T. (2008). Bundling of freight flows and hinterland network developments. In: Konings, R., Priemus, H., Nijkamp, P. (eds). *The Future of Intermodal Freight Transport*. Cheltenham: Edward Elgar, pp. 66–88.

Notteboom, T.E. (2010). Concentration and the formation of multi-port gateway regions in the European container port system: an update. *Journal of Transport Geography.* 18 (4): 567–83.

Notteboom, T., de Langen, P., Jacobs, W. (2013). Institutional plasticity and path dependence in seaports: interactions between institutions, port governance reforms and port authority routines. *Journal of Transport Geography.* 27: 26–35.

Notteboom, T.E., Rodrigue, J. (2005). Port regionalization: towards a new phase in port development. *Maritime Policy & Management*. 32 (3): 297–313.

Notteboom, T.E., Rodrigue, J.-P. (2009a). Inland terminals within North American & European Supply Chains. In: *Transport and Communications Bulletin for Asia and the Pacific No. 78: Development of Dry Ports*. New York: UNESCAP.

Notteboom, T., Rodrigue, J.-P. (2009b). The future of containerization: perspectives from maritime and inland freight distribution. *GeoJournal*. 74 (1): 7–22.

Notteboom, T., Rodrigue, J.-P. (2012). The corporate geography of global container terminal operators. *Maritime Policy & Management*. 39 (3): 249–79.

Notteboom, T.E., Winklemans, W. (2001). Structural changes in logistics: how will port authorities face the challenge? *Maritime Policy & Management*, 28 (1): 71–89.

O'Laughlin, K.A., Cooper, J.C., Cabocal, E. (1993). *Reconfiguring European Logistics Systems*. Oak Brook: CLM.

Olivier, D., Slack, B. (2006). Rethinking the port. *Environment & Planning A*. 38 (8): 1409–27.

ORR. (2011). *Rail freight sites – ORR market study*. London: ORR.

ORR. (2012). *National Rail Trends*. Available at: http://www.rail-reg.gov.uk/server/show/nav.1862 (Accessed 26 November 2012).

Overman, H.G., Winters, L.A. (2005). The port geography of UK international trade. *Environment & Planning A*. 37 (10): 1751–68.

Padilha, F., Ng, A.K.Y. (2012). The spatial evolution of dry ports in developing economies: the Brazilian experience. *Maritime Economics & Logistics*. 14 (1): 99–121.

Pallis, A.A. (2006). Institutional dynamism in EU policy-making: the evolution of the EU maritime safety policy. *Journal of European Integration*. 28 (2): 137–57.

Panayides, P.M. (2002). Economic organisation of intermodal transport. *Transport Reviews*. 22 (4): 401–14.

Panayides, P.M. (2006). Maritime logistics and global supply chains: towards a research agenda. *Maritime Economics & Logistics*. 8 (1): 3–18.

Parkhe, A. (1991). Interfirm diversity, organizational learning, and longevity in global strategic alliances. *Journal of International Business Studies*. 22 (4): 579–601.

Parola, F., Sciomachen, A. (2009). Modal split evaluation of a maritime container terminal. *Maritime Economics & Logistics*. 11 (1): 77–97.

Patton, M.Q. (2002). *Qualitative Research and Evaluation Methods*. 3rd edn. Thousand Oaks, CA: Sage.

Peck, J. (2001). Neoliberalizing states: thin policies/hard outcomes. *Progress in Human Geography*. 25 (3): 445–55.

Pemberton, S. (2000). Institutional governance, scale and transport policy – lessons from Tyne and Wear. *Journal of Transport Geography*. 8 (4): 295–308.

Perakis, A.N., Denisis, A. (2008). A survey of short sea shipping and its prospects in the USA. *Maritime Policy & Management*. 35 (6): 591–614.

Peters, N.J., Hofstetter, J.S., Hoffmann, V.H. (2011). Institutional entrepreneurship capabilities for interorganizational sustainable supply chain strategies. *The International Journal of Logistics Management*. 22 (1): 52–86.

Pettit, S.J., Beresford, A.K.C. (2009). Port development: from gateways to logistics hubs. *Maritime Policy & Management*. 36 (3): 253–67.

Pettit, S.J., Beresford, A.K.C. (2008). An assessment of long-term United Kingdom port performance: a regional perspective. *Maritime Economics & Logistics*. 10 (1): 53–74.

Pfohl, H.-C., Buse, H.P. (2000). Inter-organizational logistics systems in flexible production networks. *International Journal of Physical Distribution and Logistics Management*. 30 (5): 388–408.

Pittman, R. (2004). Chinese railway reform and competition: lessons from the experience in other countries. *Journal of Transport Economics & Policy*. 38 (2): 309–32.

Pittman, R. (2011). Risk-averse restructuring of freight railways in China. *Utilities Policy*. 19 (3): 152–60.

Port of Barcelona. (2010). *Annual Report 2009*. Barcelona: Autoridad Portuaria de Barcelona.

Port Strategy. (2012). Maersk calls ports to the table. Port Strategy. 17 October 2011. Available at: http://www.portstrategy.com/news101/products-and-services/maersk-calls-ports-to-the-table (Accessed 2 September 2013).

Porter, J. (2013). Maersk snubs Panama Canal with shift to Suez. Lloyd's List. 4 March 2013. Available at: http://www.lloydslist.com/ll/sector/containers/article417648.ece (Accessed 24 August 2013).

Potter, A., Mason, R., Lalwani, C. (2007). Analysis of factory gate pricing in the UK grocery supply chain. *International Journal of Retail and Distribution Management*. 35 (10): 821–34.

Priemus, H. (1999). On modes, nodes and networks: technological and spatial conditions for a breakthrough towards multimodal terminals and networks of freight transport in Europe. *Transportation Planning and Technology*. 23: 83–103.

Priemus, H., Zonneveld, W. (2003). What are corridors and what are the issues? Introduction to special issue: the governance of corridors. *Journal of Transport Geography*. 11 (3): 167–77.

Proost, S., Dunkerley, F., De Borger, B., Gühneman, A., Koskenoja, P., Mackie, P., Van der Loo, S. (2011). When are subsidies to trans-European network projects justified? *Transportation Research Part A*. 45 (3): 161–70.

Public Accounts Committee. (1997). *Department of Transport: Freight Facilities Grants in England, 24th Report*. London: PAC.

Puertos del Estado. (2009). *Annuario Estadístico 2008*. Madrid: Puertos del Estado.

Raco, M. (1998). Assessing 'institutional thickness' in the local context: a comparison of Cardiff and Sheffield. *Environment & Planning A*. 30 (6): 975–96.

Raco, M. (1999). Competition, collaboration and the new industrial districts: examining the institutional turn in local economic development. *Urban Studies*. 36 (5–6): 951–68.

Racunica, I., Wynter, L. (2005). Optimal location of intermodal freight hubs. *Transportation Research Part B*. 39 (5): 453–77.

Rafiqui, P.S. (2009). Evolving economic landscapes: why new institutional economics matters for economic geography. *Journal of Economic Geography*. 9 (3): 329–53.

Rahimi, M., Asef-Vaziri, A., Harrison, R. (2008). An inland port location-allocation model for a regional intermodal goods movement system. *Maritime Economics & Logistics*. 10 (4): 362–79.

Regmi, M.B., Hanaoka, S. (2012). Assessment of intermodal transport corridors: cases form north-east and central Asia. *Research in Transportation Business & Management*. 5: 27–37.

RHA. (2007). Inhibitors to the Growth of Rail Freight. Edinburgh: Scottish Executive.

Rhodes, R.A.W. (1994). The hollowing out of the state: the changing nature of the public service in Britain. *The Political Quarterly*. 65 (2): 138–51.

Richey, R.G., Roath, A.S., Whipple, J.M., Fawcett, S.E. (2010). Exploring a governance theory of supply chain management: barriers and facilitators to integration. *Journal of Business Logistics*. 31 (1): 237–56.

Rimmer, P.J. (1967). The search for spatial regularities in the development of Australian seaports 1861 – 1961/2. *Geograkiska Annaler*. 49: 42–54.

Rimmer, P.J., Comtois, C. (2009). China's container-related dynamics, 1990–2005. *GeoJournal*. 74 (1): 35–50.

Rinehart, L.M., Eckert, J.A., Handfield, R.B., Page, T.J., Atkin, T. (2004). Structuring supplier-customer relationships. *Journal of Business Logistics*. 25 (1): 25–62.

Robinson, R. (2002). Ports as elements in value-driven chain systems: the new paradigm. *Maritime Policy & Management*. 29 (3): 241–55.

Rodrigue, J.-P. (2004). Freight, gateways and mega-urban regions: the logistical integration of the Bostwash corridor. *Tijdschrift voor Economische en Sociale Geografie*. 95 (2): 147–61.

Rodrigue, J.-P. (2006). Challenging the derived transport-demand thesis: geographical issues in freight distribution. *Environment & Planning A*. 38 (8): 1449–62.

Rodrigue, J.-P. (2008). The thruport concept and transmodal rail freight distribution in North America. *Journal of Transport Geography*. 16 (4): 233–46.

Rodrigue, J.-P. (2010). *Rickenbacker Global Logistics Park, Columbus, Ohio*. Available at: http://people.hofstra.edu/geotrans/eng/ch4en/appl4en/rickenbacker.html (Accessed 22 February 2011).

Rodrigue, J.-P., Comtois, C., Slack, B. (2013). *The Geography of Transport Systems*. 3rd edn. Abingdon: Routledge.

Rodrigue, J.-P., Debrie, J., Fremont, A., Gouvernal, E. (2010). Functions and actors of inland ports: European and North American dynamics. *Journal of Transport Geography.* 18 (4): 519–29.

Rodrigue, J.-P., Notteboom, T. (2009). The terminalisation of supply chains: reassessing the role of terminals in port/hinterland logistical relationships. *Maritime Policy & Management.* 36 (2): 165–83.

Rodrigue, J.-P., Notteboom, T. (2010). Comparative North American and European gateway logistics: the regionalism of freight distribution. *Journal of Transport Geography.* 18 (4): 497–507.

Rodrigue, J.-P., Notteboom, T. (2011). Looking inside the box: evidence from the containerization of commodities and the cold chain. Paper presented at the European Conference on Shipping and Ports, Chios Greece, June 2011.

Rodrigue, J.-P, Wilmsmeier, G. (2013). *The Benefits of Logistics Investments: Opportunities for Latin America and the Caribbean.* Washington DC: Inter-American Development Bank.

Rodríguez-Pose, A. (2013). Do institutions matter for regional development? *Regional Studies.* 47 (7): 1034–47.

Rodríguez-Pose, A., Gill, N. (2003). The global trend towards devolution and its implications. *Environment & Planning C.* 21 (3): 333–51.

Roe, M. (2009). Multi-level and polycentric governance: effective policymaking for shipping. *Maritime Policy & Management.* 36 (1): 39–56.

Roe, M. (2007). Shipping, policy & multi-level governance. *Maritime Economics & Logistics.* 9 (1): 84–103.

Romein, A., Trip, J.J., de vries, J. (2003). The multi-scalar complexity of infrastructure planning: evidence from the Dutch-Flemish megacorridor. *Journal of Transport Geography.* 11 (3): 205–13.

Rong, Z., Bouf, D. (2005). How can competition be introduced into Chinese railways? *Transport Policy.* 12 (4): 345–52.

Roso, V. (2008). Factors influencing implementation of a dry port. *International Journal of Physical Distribution & Logistics Management.* 38 (10): 782–98.

Roso, V., Lumsden, K. (2010). A review of dry ports. *Maritime Economics & Logistics.* 12 (2): 196–213.

Roso, V., Woxenius, J., Lumsden, K. (2009). The dry port concept: connecting container seaports with the hinterland. *Journal of Transport Geography.* 17 (5): 338–45.

Roson, R., Soriani, S. (2000). Intermodality and the changing role of nodes in transport networks. *Transportation Planning and Technology.* 23: 183–97.

RTI. (2000). *Transportation and the Potential for Intermodal Efficiency Enhancements in Western West Virginia.* Report prepared by the Nick J. Rahall Appalachian Transportation Institute on behalf of the Appalachian Regional Commission, the West Virginia DOT and West Virginia Planning and Regional Development Council. Huntington: RTI.

RTI. (2003). *Central Corridor Double-Stack Initiative: Final Report.* Report prepared by the Nick J. Rahall Appalachian Transportation Institute. Huntington: RTI.

Runhaar, H., van der Heijden, R. (2005). Public policy intervention in freight transport costs: effects on printed media logistics in the Netherlands. *Transport Policy.* 12 (1): 35–46.

Sánchez, R., Wilmsmeier, G. (2010). Contextual Port Development: A Theoretical Approach. In: Coto-Millán, P., Pesquera, M., Castanedo, J. (eds). *Essays on Port Economics.* New York: Springer, pp. 19–44.

Schaetzl, L. (1996). *Wirtschaftsgeographie 1 – Theorie.* 6th edn. Paderborn: UTB.

Schmoltzi, C., Wallenburg, C.M. (2011). Horizontal cooperations between logistics service providers: motives, structure, performance. *International Journal of Physical Distribution and Logistics Management.* 41 (6): 552–76.

Schober, H. (1991). Irritation und Bestätigung – Die Provokation der systemischen Beratung oder: Wer macht eigentlich die Veränderung? In: Hofmann, M. (ed.). *Theorie und Praxis der Unternehmensberatung.* Heidelberg: Physica, pp. 345–70.

Scott, W.R. (2008). *Institutions and Organizations.* 3rd edn. Los Angeles: Sage.

Scott, W.R., Meyer, J.W. (1983). The organization of societal sectors. In: Meyer, J.W., Scott, W.R. (eds). *Organizational Environments: Ritual and Rationality.* Beverly Hills, CA: Sage, pp. 129–53.

Scottish Government. (2010). *Food and Drink in Scotland: Key Facts 2010.* Edinburgh: Scottish Government.

Seale, C. (1999). *The Quality of Qualitative Research.* London: Sage.

Shaw, J., Sidaway, J.D. (2010). Making links: On (re)engaging with transport and transport geography. *Progress in Human Geography.* 35 (4): 502–20.

Simatupang, T.M., Sridharan, R. (2005). The collaboration index: a measure for supply chain collaboration. *International Journal of Physical Distribution and Logistics Management.* 35 (1): 44–62.

Simons, H. (2009). *Case Study Research in Practice.* London: Sage.

Skjoett-Larsen, T. (2000). Third party logistics – from an interorganizational point of view. *International Journal of Physical Distribution and Logistics Management.* 30 (2): 112–27.

Slack, B. (1990). Intermodal transportation in North America and the development of inland load centres. *Professional Geographer.* 42 (1): 72–83.

Slack, B. (1993). Pawns in the game: ports in a global transport system. *Growth and Change.* 24 (4): 579–88.

Slack, B. (1999). Satellite terminals: a local solution to hub congestion? *Journal of Transport Geography.* 7 (4): 241–6.

Slack, B. (2007). The terminalisation of seaports. In: Wang, J., Olivier, D., Notteboom, T., Slack, B. (eds). *Ports, Cities and Global Supply Chains.* Aldershot: Ashgate, pp. 41–50.

Slack, B., Frémont, A. (2005). Transformation of port terminal operations: from the local to the global. *Transport Reviews.* 25 (1): 117–30.

Slack, B., Vogt, A. (2007). Challenges confronting new traction providers of rail freight in Germany. *Transport Policy.* 14 (5): 399–409.

Slack, B., Wang, J.J. (2002). The challenge of peripheral ports: an Asian perspective. *GeoJournal*. 56 (2): 159–66.

Smith, D.L.G., Sparks, L. (2004). Logistics in Tesco: past, present and future. In: Fernie, J., Sparks, L. (eds). *Logistics and Retail Management*. 2nd edn. London: Kogan Page, pp. 101–20.

Smith, D.L.G., Sparks, L. (2009). Tesco's supply chain management. In: Fernie, J., Sparks, L. (eds). *Logistics and Retail Management*. 3rd edn. London: Kogan Page, pp. 143–71.

Song, D.-W., Panayides, P.M. (2008). Global supply chain and port/terminal: integration and competitiveness. *Maritime Policy & Management*. 35 (1): 73–87.

Spekman R.E., Kamauff, J.W., Myhr, N. (1998). An empirical investigation into supply chain management: a perspective on partnerships. *International Journal of Physical Distribution & Logistics Management*. 28 (8): 630–50.

SRA. (2004). *Strategic Rail Freight Interchange Policy*. London: SRA.

Stake, R.E. (1995). *The Art of Case Study Research*. Thousand Oaks, CA: Sage.

Stank, T.P., Keller, S.B., Daugherty, P.J. (2001). Supply chain collaboration and logistical service performance. *Journal of Business Logistics*. 22 (1): 29–48.

Steinberg P. (2001). *The Social Construction of the Ocean*. Cambridge: Cambridge University Press.

Stone, J.I. (2001). *Infrastructure Development in Landlocked and Transit Developing Countries: Foreign Aid, Private Investment and the Transport Cost Burden of Landlocked Developing Countries*. UNCTAD\LDC\112. Geneva: UNCTAD.

Storper, M. (1997). *The Regional World: Territorial Development in a Global Economy*. New York: Guilford Press.

Stough, R.R., Rietveld, P. (1997). Institutional issues in transport systems. *Journal of Transport Geography*. 5 (3): 207–14.

Strambach, S. (2010). Path dependency and path plasticity: the co-evolution of institutions and innovation – the German customized business software industry. In: Boschma, R., Martin, R. (eds). *Handbook of Evolutionary Economic Geography*. Cheltenham, Edward Elgar, pp. 406–31.

Suchman, M.C. (1995). Managing legitimacy: strategic and institutional approaches. *Academy of Management Review*. 20 (3): 571–610.

Swyngedouw, E. (1992). Territorial organization and the space/technology nexus. *Transactions of the Institute of British Geographers*. 17 (4): 417–33.

Swyngedouw, E. (1997). Neither global nor local: 'Glocalisation' and the politics of scale. In: Cox, K. (ed.). *Spaces of Globalization*. New York: Guildford, pp. 137–66.

Swyngedouw, E. (2000). Authoritarian governance, power and the politics of rescaling. *Environment & Planning D*. 18 (1): 63–76.

Taaffe, E.J., Morrill, R.L., Gould, P.R. (1963). Transport expansion in underdeveloped countries: a comparative analysis. *Geographical Review*. 53: 503–29.

Talley, W.K. (2009). *Port Economics*. Abingdon: Routledge.

Todeva, E., Knoke, D. (2005). Strategic alliances and models of collaboration. *Management Decision*. 43 (1): 123–48.

Towill, D.R. (2005). A perspective on UK supermarket pressures on the supply chain. *European Management Journal*. 23 (4): 426–38.

Trip, J., Bontekoning, Y. (2002). Integration of small freight flows in the intermodal transport system. *Journal of Transport Geography*. 10 (3): 221–9.

Tsamboulas, D.A., Kapros, S. (2003). Freight village evaluation under uncertainty with public and private financing. *Transport Policy*. 10 (2): 141–56.

Tsamboulas, D., Vrenken, H., Lekka, A.-M. (2007). Assessment of a transport policy potential for intermodal mode shift on a European scale. *Transportation Research Part A*. 41 (8): 715–33.

UIR. (2009). *Il Disegno Dell'interportualita Italiana*. Milan: FrancoAngeli.

UNCTAD. (1982). *Multimodal Transport and Containerization*. TD/B/C.4/238/ supplement 1, Part Five: Ports and Container Depots. Geneva: UNCTAD.

UNCTAD. (1992). *Development and Improvement of Ports: the Principles of Modern Port Management and Organisation*. Geneva: UNCTAD.

UNCTAD. (2004). *Assessment of a Seaport Land Interface: an Analytical Framework*. Geneva: UNCTAD.

UNCTAD. (2013). *The Way to the Ocean; Transit Corridors Servicing the Trade of Landlocked Developing Countries*. Geneva: UNCTAD.

UNESCAP. (2006). *Cross-Cutting Issues for Managing Globalization Related to Trade and Transport: Promoting Dry Ports as a Means of Sharing the Benefits of Globalization with Inland Locations*. Bangkok, Thailand: UNESCAP.

UNESCAP. (2008). *Policy Framework for the Development of Intermodal Interfaces as Part of an Integrated Transport Network in Asia*. Bangkok, Thailand: UNESCAP.

Van de Voorde, E., Vanelslander, T. (2009). *Market power and vertical and horizontal integration in the maritime shipping and port industry*. JTRC OECD/ITF Discussion Paper 2009–2. Paris: ITF.

Van den Berg, R., de Langen, P.W. (2011). Hinterland strategies of port authorities: a case study of the Port of Barcelona. *Research in Transportation Economics*. 33: 6–14.

Van den Berg, R., De Langen, P.W., Costa, C.R. (2012). The role of port authorities in new intermodal service development: the case of Barcelona Port Authority. *Research in Transportation Business & Management*. 5: 78–84.

Van der Horst, M.R., Van der Lugt, L.M. (2009). Coordination in railway hinterland chains: an institutional analysis. Paper presented at the annual conference of the International Association of Maritime Economists (IAME), Copenhagen, June 2009.

Van der Horst, M.R., De Langen, P.W. (2008). Coordination in hinterland transport-chains: a major challenge for the seaport community. *Maritime Economics & Logistics*. 10 (1–2): 108–29.

Van der Horst, M.R., Van der Lugt, L.M. (2011). Coordination mechanisms in improving hinterland accessibility: empirical analysis in the port of Rotterdam. *Maritime Policy & Management*. 38 (4): 415–35.

Van Ham, H., Rijsenbrij, J. (2012). *Development of Containerization: Success Through Vision, Drive and Technology*. Amsterdam: IOS Press.

Van Ierland, E., Graveland, C., Huiberts, R. (2000). An environmental economic analysis of the new rail link to European main port Rotterdam. *Transportation Research Part D*. 5 (3): 197–209.

Van Klink, H.A. (1998). The port network as a new stage in port development: the case of Rotterdam. *Environment and Planning A*. 30 (1): 143–60.

Van Klink, H.A. (2000). Optimisation of land access to sea ports. In: *Land Access to Sea Ports, European Conference of Ministers of Transport*. Paris, December 1998, pp. 121–41.

van Klink, H.A., van den Berg, G.C. (1998). Gateways & Intermodalism. *Journal of Transport Geography*. 6 (1): 1–9.

Van Schijndel, W.J., Dinwoodie, J. (2000). Congestion and multimodal transport: a survey of cargo transport operators in the Netherlands. *Transport Policy*. 7 (4): 231–41.

Van Schuylenburg, M., Borsodi, L. (2010). Container transferium: an innovative logistic concept. Available at:http://www.citg.tudelft.nl/fileadmin/Faculteit/CiTG/Over_de_faculteit/Afdelingen/Afdeling_Waterbouwkunde/sectie_waterbouwkunde/chairs/ports_and_waterway/Port_Seminar_2010/Papers_and_presentations/doc/Paper_Schuylenburg_-_Container_Transferium_Port_Seminar__2_.pdf (Accessed 9 April 2013).

Veenstra, A., Zuidwijk, R., Van Asperen, E. (2012). The extended gate concept for container terminals: expanding the notion of dry ports. *Maritime Economics and Logistics*. 14 (1): 14–32.

Verhoeven, P. (2009). European ports policy: meeting contemporary governance challenges. *Maritime Policy & Management*. 36 (1): 79–101.

Verhoeven, P., Vanoutrive, T. (2012). A quantitative analysis of European port governance. *Maritime Economics & Logistics*. 14 (2): 178–203.

Von Thünen, J.H. (1826). *Der Isolierte Staat (The Isolated State)*. Trans. C.M. Wartenberg (1966). Oxford: Pergamon.

Wackett, M. (2013). Carriers rethink one-stop-shop strategy. Lloyd's List. 22 August 2013. Available at: http://www.lloydslist.com/ll/sector/containers/article428237.ece (Accessed 3 September 2013).

Wang, C., Ducruet, C. (2012). New port development and global city making: emergence of the Shanghai-Yangshan multi-layered gateway hub. *Journal of Transport Geography*. 25: 58–69.

Wang, J.J., Ng, A.K.Y., Olivier, D. (2004). Port governance in China: a review of policies in an era of internationalizing port management practices. *Transport Policy*. 11 (3): 237–50.

Wang, J., Olivier, D., Notteboom, T., Slack, B. (eds). (2007). *Ports, Cities and Global Supply Chains*. Aldershot: Ashgate.

Wang, J.J., Slack, B. (2004). Regional governance of port development in China: a case study of Shanghai International Shipping Centre. *Maritime Policy & Management*. 31 (4): 357–73.

Wang, K., Ng, A.K.Y., Lam, J.S.L., Fu, X. (2012). Cooperation or competition? Factors and conditions affecting regional port governance in South China. *Maritime Economics & Logistics*. 14 (3): 386–408.

Wathne, K.H., Heide, J.B. (2000). Opportunism in interfirm relationships: forms, outcomes, and solutions. *Journal of Marketing*. 64 (4): 36–51.

Weber, A. (1909). *Über den Standort der Industrien (Theory of the Location of Industries)*. Trans. C.J. Friedrich (1929). Chicago: The University of Chicago Press.

Wernerfelt, B. (1984). A resource-based view of the firm. *Strategic Management Journal*. 5 (2): 171–80.

Whipple, J.M., Frankel, R. (2000). Strategic alliance success factors. *The Journal of Supply Chain Management*. 36 (3): 21–8.

Whipple, J.M., Russell, D. (2007). Building supply chain collaboration: a typology of collaborative approaches. *The International Journal of Logistics Management*. 18 (2): 174–96.

Wiegmans, B.W., Masurel, E., Nijkamp, P. (1999). Intermodal freight terminals: an analysis of the terminal market. *Transportation Planning & Technology*. 23 (2): 105–28.

Wilding, R., Humphries, A.S. (2006). Understanding collaborative supply chain relationships through the application of the Williamson organisational failure framework. *International Journal of Physical Distribution and Logistics Management*. 36 (4): 309–29.

Williamson, O.E. (1975). *Markets and Hierarchies: Analysis and Antitrust Implications*. New York: The Free Press.

Williamson, O.E. (1985). *The Economic Institutions of Capitalism*. New York: The Free Press.

Wilmsmeier, G., Monios, J. (2013). Counterbalancing peripherality and concentration: an analysis of the UK container port system. *Maritime Policy & Management*. 40 (2): 116–32.

Wilmsmeier, G., Monios, J., Lambert, B. (2010). Observations on the regulation of "dry ports" by national governments. Paper presented at the annual conference of the International Association of Maritime Economists (IAME), Lisbon, July 2010.

Wilmsmeier, G., Monios, J., Lambert, B. (2011). The directional development of intermodal freight corridors in relation to inland terminals. *Journal of Transport Geography*. 19 (6): 1379–86.

Wilmsmeier, G., Monios, J., Perez, G. (2013). Port system evolution – the case of Latin America and the Caribbean. Paper presented at the annual conference of the International Association of Maritime Economists (IAME). Marseille, France, July 2013.

Wilmsmeier, G., Notteboom, T. (2011). Determinants of liner shipping network configuration: a two-region comparison. *GeoJournal*. 76 (3): 213–28.

Woo, S.-H., Pettit, S.J., Kwak, D.-W., Beresford, A.K.C. (2011a). Seaport research: a structured literature review on methodological issues since the 1980s. *Transportation Research Part A*. 45 (7): 667–85.

Woo, S.-H., Pettit, S., Beresford, A.K.C. (2011b). Port evolution and performance in changing logistics environments. *Maritime Economics & Logistics*. 13 (3): 250–77.

Woodburn, A. (2003). A logistical perspective on the potential for modal shift of freight from road to rail in Great Britain. *International Journal of Transport Management*. 1 (4): 237–45.

Woodburn, A. (2007). Evaluation of rail freight facilities grants funding in Britain. *Transport Reviews*. 27 (3): 311–26.

Woodburn, A. (2008). Intermodal rail freight in Britain: a terminal problem? *Planning, Practice & Research*. 23 (3): 441–60.

Woodburn, A. (2011). An investigation of container train service provision and load factors in Great Britain. *European Journal of Transport and Infrastructure Research*. 11 (2): 147–65.

Woodburn, A. (2012). Intermodal rail freight activity in Britain: where has the growth come from? *Research in Transportation Business & Management*. 5: 16–26.

World Bank. (2001). *Port Reform Toolkit*. Washington DC: World Bank.

World Bank. (2007). *Port Reform Toolkit*. 2nd edn. Washington DC: World Bank.

Woxenius, J., Bärthel, F. (2008). Intermodal road-rail transport in the European Union. In: Konings, R., Priemus, H., Nijkamp, P. (eds). *The Future of Intermodal Freight Transport*. Cheltenham: Edward Elgar, pp. 13–33.

Woxenius, J., Bergqvist, R. (2011). Comparing maritime containers and semi-trailers in the context of hinterland transport by rail. *Journal of Transport Geography*. 19 (4): 680–88.

WSP. (2006). *Scottish Freight Strategy Scoping Study*. Report prepared for the Scottish Executive. Edinburgh: Scottish Executive.

Wu, J.H., Nash, C. (2000). Railway reform in China. *Transport Reviews*. 29 (1): 25–48.

Xie, R., Chen, H., Nash, C. (2002). Migration of railway freight transport from command economy to market economy: the case of China. *Transport Reviews*. 22 (2): 159–77.

Yin, R. (2009). *Case Study Research*. Thousand Oaks, CA: Sage.

Yin, R. (2012). *Applications of Case Study Research*. Thousand Oaks, CA: Sage.

Zacharia, Z.G., Sanders, N.R., Nix, N.W. (2011). The emerging role of the third-party logistics provider (3PL) as an orchestrator. *Journal of Business Logistics*. 32 (1): 40–54.

Index

For Product Safety Concerns and Information please contact our EU representative GPSR@taylorandfrancis.com Taylor & Francis Verlag GmbH, Kaufingerstraße 24, 80331 München, Germany

Printed and bound by CPI Group (UK) Ltd, Croydon, CR0 4YY

08/05/2025

01864338-0002